GORILLAS

GORILLAS

SARA GODWIN

PRINCIPAL PHOTOGRAPHY BY GERRY ELLIS

MALLARD PRESS

MALLARD PRESS

An imprint of BDD Promotional Book Company, Inc.
666 Fifth Avenue
New York, New York 10103

A FRIEDMAN GROUP BOOK

Published by MALLARD PRESS
An imprint of BDD Promotional Book Company, Inc.
666 Fifth Avenue
New York, New York 10103

Mallard Press and its accompanying design and logo are trademarks
of BDD Promotional Book Company, Inc.

ISBN 0-792-45259-3

GORILLAS
was prepared and produced by
Michael Friedman Publishing Group, Inc.
15 West 26th Street
New York, New York 10010

Editor: Melissa Schwarz
Art Director: Jeff Batzli
Designer: Lynne Yeamans
Photography Editor: Christopher Bain
Production: Karen L. Greenberg

All contemporary photographs not credited elsewhere © Gerry Ellis.

Typeset by: BPE Graphics, Inc.
Color separations by Universal Colour Scanning, Ltd.
Printed and bound in Hong Kong by Leefung-Asco Printers, Ltd.

DEDICATION

For C. J., by my side, on my side, through all the lovely years

ACKNOWLEDGMENTS

I must extend my grateful thanks to Carol Martinez, Senior Primate
Keeper of the San Francisco Zoo, who allowed me to visit
Gorilla World with its two babies, Binti Jua and Shango; to The Gorilla
Foundation which kept me abreast of Koko's extraordinary
achievements; to the National Geographic Society which has
generously supported and reported gorilla research for many years;
and to Chuck James, who reviewed much of the current
scientific literature available on gorillas and generously gave me the
benefit of his insights. Any errors, God forbid, are exclusively my own.

C O N T

E N T S

CHAPTER 1

*Myth,
Legend,
and
History*

If ever a creature suffered from a bad reputation, surely it is the gorilla. Acknowledged as the greatest of man's closest kin, the great apes, the gorilla has been vilified and feared as a ferocious, terrifying beast capable of tearing a man limb from limb and reputed to steal native women and carry them off. Only recently has the true nature of the gorilla emerged: like humans in its intelligence, like humans in its depth of feeling, and utterly unlike humans in its gentleness. Dr. George Schaller, director for science of Wildlife Conservation International (a division of the New York Zoological Society) and a leading authority on the gorilla, characterizes most of the published information on this much-maligned creature as "sensational, irresponsible, exaggerated prevarications, with little concern for truth." As thoroughly as *Homo sapiens* as a species has earned James Joyce's painfully accurate description of "manunkind", so *Gorilla gorilla gorilla* deserves the title "gentleman" in a way few humans can honestly claim.

In the mid-1800's, killing a gorilla was considered an act of great skill and courage.

King Kong, *released in 1933, shaped the nation's perception of gorillas for fifty years. King Kong killed sacrificial virgins, fought pterodactyls and other prehistoric monsters, climbed the Empire State Building with one hand (clutching Fay Wray in the other), and snatched armed biplanes out of the air.*

Never intended as anything more than a science fiction fantasy, **King Kong** *made a profound impression on the public mind that required the combined efforts of Dian Fossey's fourteen years of research, several* **National Geographic** *articles, and the movie,* **Gorillas in the Mist,** *to eradicate.*

FPG International

King Kong was a fantasy gorilla invented only to terrify; Koko is a real gorilla who has learned to speak and read. Gargantua was the ferocious-looking star attraction of the prewar Barnum & Bailey circus, while big-eyed baby Patty Cake of New York's Central Park Zoo demonstrated unmistakably the close bonding between gorilla mother and child. *Congorilla* was the first film ever made of mountain gorillas in the wild; *Gorillas in the Mist* is the story of Dian Fossey's fourteen years of research with the mountain gorilla that ended in her murder by machete. How did we get from King Kong to Koko, from Gargantua to Patty Cake, from Martin and Osa Johnson's *Congorilla* to Warner Brothers' *Gorillas in the Mist*?

The story begins in the nineteenth century when the western European powers were carving up Africa to suit their own tastes. Colonization was carried out by missionaries sent to bring the heathen to Christ and by entrepreneurs sent to wrest the raw materials from the land to supply the factories of the Industrial Revolution. As adventurers and explorers penetrated the heart of the so-called Dark Continent, they sent back reports of the wonders they saw. Rumors of the gorilla, "a wonderful production of nature [who] walks upright like a man; is from 7 to 9 feet (210 to 270 cm) high, when at maturity, thick in proportion, and amazingly strong" began filtering into Europe as early as 1625, but the credit for sending back the first skulls for proper scientific classification belongs to an American missionary, Reverend J. L. Wilson of the American Board of Foreign Missions to West Africa, who obtained skulls from the natives of Gabon in 1846. These were turned over to Dr. Thomas S. Savage of the Boston Society of Natural History. In 1847 Savage sent the skulls to anatomists Jeffries Wyman and Richard Owen at Harvard University, accompanied by a description remarkable mainly for its inaccuracy. The description opens, "They [gorillas] are exceedingly ferocious, and always offensive in their habits. . . ." Savage went on to accuse gorillas of habitually crushing gun barrels between their teeth and of killing hunters. It's rather a shame it wasn't true, for had it been, the gorilla would be less endangered today.

The anatomist Owen, who had classified and named the gorilla without ever having seen one alive, in 1859 reported a tale so lurid that Schaller refers to it as "abandoning science for mythology." Owen wrote, "Negroes when stealing through shades of the tropical forest become sometimes aware of the proximity of one of these frightfully formidable beasts by the sudden disappearance of one of their companions, who is hoisted up into the tree, uttering, perhaps, a short choking cry. In a few minutes he falls to the ground a strangled corpse."

The American explorer Paul du Chaillu distinguished himself in 1856 by becoming the first white man to shoot a gorilla. He published the lurid account of this exploit in 1861 in his wildly popular *Explorations and Adventures in Equatorial Africa*. With these words du Chaillu established the gorilla as a ferocious beast: "And now truly he reminded me of nothing but some hellish dream creature—a being of that hideous order, half-man, half-beast, which we find pictured by old artists in some representations of the infernal regions. He advanced a few steps—then stopped to utter that hideous roar again—advanced again, and finally stopped when at a distance of about six yards from us. And here, just as he began another of his roars, beating his chest in rage, we fired, and killed him."

Despite his sensationalism, du Chaillu's account of gorilla behavior was essentially accurate, the most reliable description available for nearly a century. Carl Akeley, who single-handedly persuaded the Belgian government to establish a permanent sanctuary for the mountain gorillas in its African territory in 1925, observed in his book *In Brightest Africa*, "If you read the tale as du Chaillu wrote it, it gives the impression that the gorilla is a terrible animal. If you read merely what the gorilla did, you will see that he did nothing that a domestic dog might not do under the same circumstances." Du Chaillu's description of gorilla behavior was rejected out of hand by the scientists of the day, who preferred the man-killing version promoted by such people as the anatomist Richard Owen. Du Chaillu, of course, had a telling advantage over the scientists: He had seen gorillas, and they had not. In all fairness to du Chaillu, it is said that his publisher rejected his original manuscript, complaining that it was not sufficiently lively, which may explain some of the exaggeration and melodrama of the published version. Even today, 128 years later, there is astonishingly little firsthand information on the behavior of the lowland gorilla in the wild.

With each repetition of stories like du Chaillu's, the gorilla grew fiercer, more powerful, and more aggressive. P. T. Barnum—the great American showman with his traveling three-ring circus,

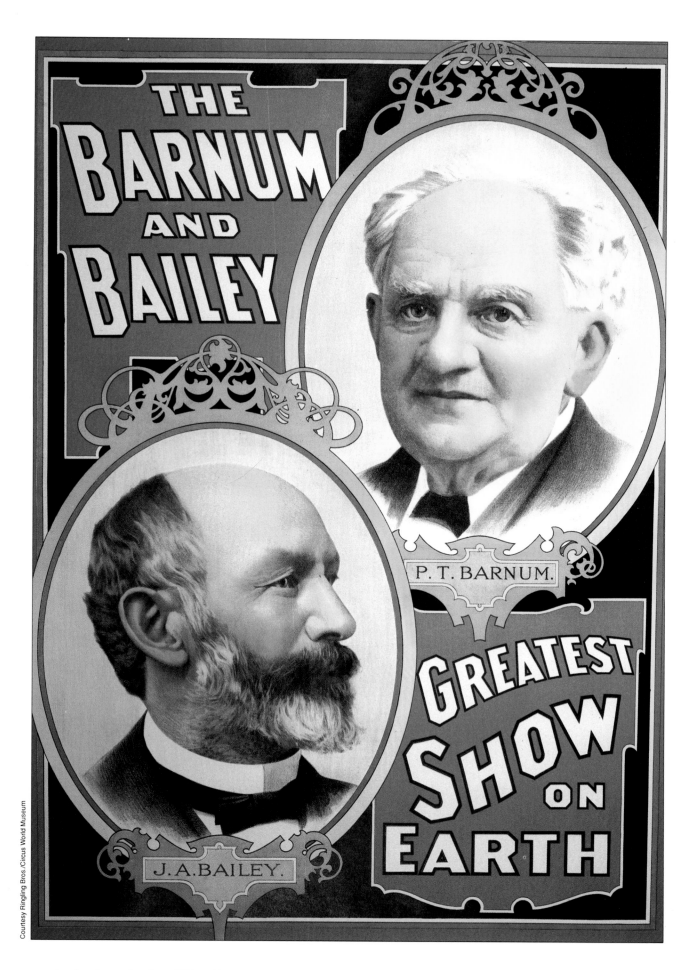

P.T. Barnum originated the idea of the three ring circus. Nineteenth-century European circuses typically consisted of clowns, magicians, jugglers, trapeze and tightrope artists, freaks, and sometimes, a dancing bear. It took Barnum to put it all together under one tent, with lion tamers, bareback riders, and acrobats, all plying their trades simultaneously. Billed as "The Greatest Show on Earth," the Barnum & Bailey circus desperately wanted the most terrifying creature known to man: They wanted a gorilla. Barnum funded his own expeditions to get one.

modestly described as "The Greatest Show on Earth"—wrote in 1888 that, "My good friend du Chaillu tells me he made repeated attempts to win the regard of the young [gorillas] he was fortunate enough to capture, but was never able to awaken the first spark of affection in them." The remark appears in Barnum's *The Wild Beasts, Birds, and Reptiles of the World: The Story of Their Capture*, a rousing adventure tale of how the animals were secured for his famous circus and sideshows. Given Barnum's occupation, one is inclined to suspect his account was painted liberally with the thick brush of embellishment.

By this time, exhibiting a gorilla was a particular prize, since none had yet survived in captivity in the United States—and Barnum wanted one desperately for his circus. Convinced that an adult gorilla was too large to bring back alive and less likely to be trained to do circus tricks, Barnum's wild animal hunters were determined to capture a baby gorilla. Three chapters of his book are devoted to the tale of how they did it. Within minutes, the hunters watch one gorilla attack and subdue a massive crocodile, see another destroy an attacking leopard with a single bite, and shoot a third that has killed one of their African guides with a single blow after bending the barrel of his gun nearly in half.

The crocodile and the gorilla are two of the most ferocious members of the animal kingdom. Both stand low on the plane of intelligence, but each possesses a certain cunning which enables him to use his prodigious strength to the best advantage. . . . The gorilla . . . walked lightly toward the crocodile, as if he intended to attack in front. Uttering his cry in a half-suppressed voice the gorilla made a leap forward, as if to alight on the snout of the other. Instantly those jaws opened like a vast steel trap, and, had the gorilla made the bound that he appeared to have started upon, he would have been caught in a vise from which ten times his power would not have extricated him. But, with inimitable dexterity, the animal turned himself to one side and leaped backward, eluding the mouth, which snapped shut with a sound that startled the spectators. He bounded into the air with a nimbleness that could not have been surpassed, and the next moment did a thing so incredible that the hunters could hardly believe their eyes. The gorilla darted like a flash to the left, then sprang directly upon the back of the saurian, and, bending over, grasped his forelegs. One was seized with either hand, and, summoning his Samson-like strength, he leaned backward and jerked with might and main. The spectators heard the bones crack, and they knew that both the crocodile's legs were broken like pipestems. The reptile struggled fiercely, lashing his tail, contorting his body, and snapping his jaws in a way that seemed impossible in one so grieviously wounded. . . .

Meanwhile, Bob Marshall detected a handsome female leopard a short distance in front of him, whose actions showed she had her attention fixed upon something so far in advance that it was invisible to him. He was still wondering what her intended victim could be, when he saw something stirring in the undergrowth just beyond. He could not distinguish it clearly, and was still trying to do so when his blood was set tingling by a wild, resounding "kh-ah! kh-ah!" the well-known cry of the gorilla. At the same moment, the dimly-seen object in advance of the leopard resolved itself into a female of the [gorilla] species, which dashed off among the limbs, fallen trees, and running vines with her young one held to her breast, as a mother clasps her baby. The leopard had hardly time to rise to her feet with the purpose of dashing after the female gorilla and her young, when the male, a gigantic fellow, fully six feet tall, burst through the vegetation, and assailed her with inconceivable ferocity. The astounded Bob saw the beam-like arms make a terrific sweep through the air, and in an instant the leopard was grasped and flung on its side. One of the gorilla's enormous arms gripped his prey under the throat, while the other, passing over her shoulders, seized the left paw and held it as immovable as if it were the

GORILLAS

16

arm of an infant. This brought the leopard's head under the chin of the terrible creature, which opened its vast jaws until they inclosed [sic] half the leopard's neck between them. Then the teeth met, and the victim had barely time to give one frenzied screech, when her life went out like a flash of lightning, and that, too, before she was able to inflict so much as a scratch upon her fearful assailant. Suddenly he flung up his head, half straightened his body, and gave the carcass a fling which sent it flying, end over end, among the branches to a point fifty feet away.

Hargo [one of the African stalkers] had never seen such a formidable creature, which, instead of fleeing, appeared to be eager for a fight. Walking forward, the gorilla stopped a rod away, struck his tremendous breast with his paw, sending out a sound like a bass drum, and, opening his vast crimson mouth to its full extent, emitted a roar that was enough to test the nerves of the bravest man. Then he commenced walking slowly forward, with his wicked black eyes fixed on the native and the hair over his skull twitching with rage. Knowing he had but one shot, Hargo held his fire until the gorilla had advanced within ten paces, and with the muzzle of his weapon almost within reach of his paw, pulled the trigger. Alas! the piece missed fire, and the instant it did so the African knew he was doomed. He gave a sweeping blow at the monster with his gun. The blow landed, but produced no more effect than if it had struck the side of a tree. In an instant the weapon was wrenched from his grasp by a single blow of the gorilla, who smote the man with the other, the blow crushing his skull as if it had been cardboard. The beast still held the gun, and as if in the pure wantonness of strength, he bent it over in a half circle, apparently with no more effort than if it were a thin pipe of lead.

(Far right) The first gorilla Dian Fossey saw on Mt. Mikeno was a lone male gorilla sunbathing on a horizontal tree trunk that projected over a small lake nestled in a corner of the Kabara meadow. When she began tracking gorilla groups she climbed trees to observe them as George Schaller had. Later, she learned that she was more successful when she stayed on the ground and let the gorillas climb trees to observe her, until they became accustomed to her presence.

Avidly read by a curious and fascinated American public, these tales did nothing to promote the reputation of the gorilla as shy, gentle, intelligent, and affectionate.

To capture one baby gorilla, Barnum's hunters killed three adult gorillas. Then, the baby, like so many before it, died before it ever reached America. In the overwrought story related above, one thing has the ring of truth: Gorillas live in tightly knit social groups, and, like people, they protect their babies. Virtually the only way to capture a young gorilla is to kill the defending adults.

There are three species of gorilla, but the differences between them are so minor that even a trained anatomist would be challenged to distinguish between them, given only a skull of each species. The first gorilla to be identified—*Gorilla gorilla gorilla*—is the western lowland gorilla. Dr. Dian Fossey estimated that there were some 10,000 western lowland gorillas living in equatorial West Africa (southern Nigeria, Gabon, Cameroon, Congo, and the Central African Republic) in 1983. The eastern lowland gorilla—*Gorilla gorilla graueri*—is found primarily in eastern Zaire, where their population numbers approximately 4,000. The mountain gorilla—*Gorilla gorilla berengi*—is seriously endangered, with a population of only about 275 in the upper elevations of the Virunga Mountains of Rwanda, Uganda, and Zaire; numbers outside that region have not been counted since 1960. The total estimated number of all free-living gorillas in the world is barely 14,275, scarcely enough to constitute a small town. There are no mountain gorillas living in captivity, and no more than two dozen captive eastern lowland gorillas. Most of the gorillas in captivity throughout the world are western lowland gorillas.

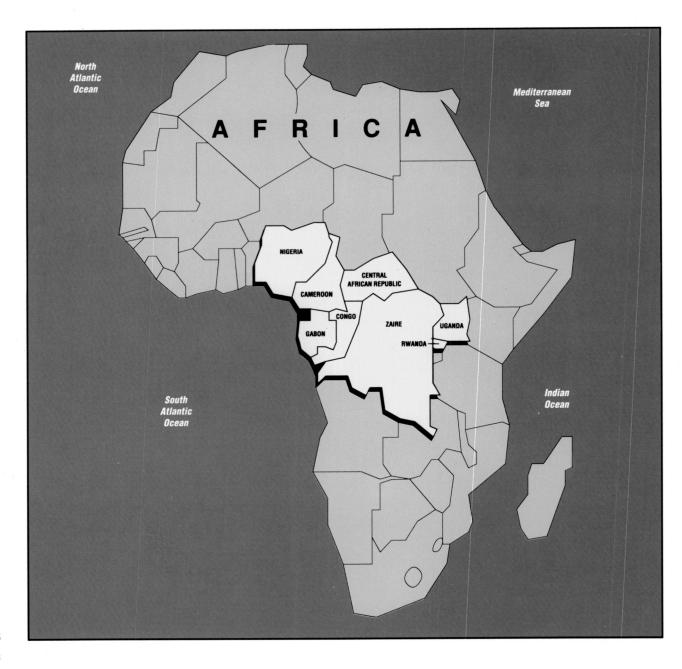

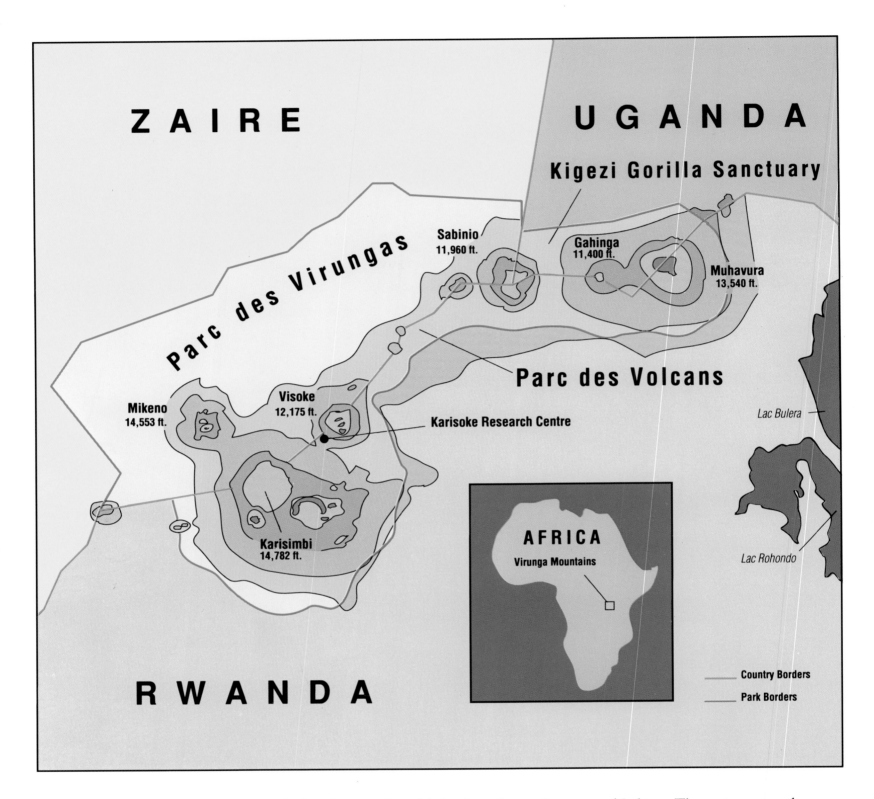

ZAIRE

UGANDA

Kigezi Gorilla Sanctuary

Parc des Virungas

Sabinio
11,960 ft.

Gahinga
11,400 ft.

Muhavura
13,540 ft.

Parc des Volcans

Mikeno
14,553 ft.

Visoke
12,175 ft.

Karisoke Research Centre

Lac Bulera

Karisimbi
14,782 ft.

AFRICA
Virunga Mountains

Lac Rohondo

RWANDA

Country Borders
Park Borders

Relatively little is known about lowland gorillas in their natural habitat. The major research on wild gorillas concerns the mountain gorilla and was conducted by George Schaller and Dian Fossey, two American scientists who originally conducted census counts to determine their numbers. Schaller began his census in 1961, Fossey in 1966; in both cases the gentle, accepting nature of the animals allowed the scientists to pursue their studies far beyond mere population counts, giving us unprecedented insight into the greatest and gentlest of the apes.

The very first suggestion that gorillas might have been misrepresented came from Carl Akeley, the famous naturalist and sculptor. In 1921 he was sent to the Belgian Congo by the American Museum of Natural History to kill five gorillas, in order to complete the gorilla exhibit at the museum. Akeley was so taken with his quarry that he persuaded the Belgian government to establish a sanctuary for the gorillas so that they might never be hunted again. Albert National Park, named for the Belgian king, was established in 1925 and enlarged in 1929. (With the withdrawal of the European powers from Africa, the gorilla conservation area became Parc National des Virungas in Zaire, the Parc National des Volcans in Rwanda, and the Kigezi Gorilla Sanctuary in Uganda.) Akeley was also the inventor of still and movie cameras and lenses specifically designed for wildlife photography. On an expedition to make the first film ever made of gorillas in the wild, Akeley died and was buried in the park he had helped create. His work—photographing and filming mountain gorillas—was carried on by Martin and Osa Johnson, explorers and filmmakers who presented the mountain gorilla to the American public in their movie *Congorilla*. They also presented the San Diego Zoo with a pair of young mountain gorillas they had captured, the first live members of the species to come to the United States. The year was 1931. They received $15,000 dollars for the pair. Today a pair of young mountain gorillas would be beyond price. A single captive-bred low-land gorilla today can cost a zoo on the order of $100,000.

Congorilla gave the western world its first view of gorillas in their native habitat—of their affectionate family groups, their strict vegetarianism, their brave defense of their little ones, and of the charges, roars, and chest-beating that are meant to frighten off, not harm, intruders. It was also the first picture with sound ever made in Central Africa. Getting anything on film at all was terribly difficult. The weather on Mt. Mikeno, where the Johnsons first tried to shoot, was either cold and rainy or cold and misty, with light too dim for filming. The sound equipment was delicate and needed constant adjustment and repair. To reach the gorillas, the Johnsons hauled their heavy movie cameras, with even heavier wooden tripods, and the fragile sound equipment up steep, slippery slopes and through stinging nettles, hacking their way into the thick jungle, all the while trying to do it silently so they wouldn't frighten off the shy gorillas, which often fled before them. They were successful in filming gorillas feeding, a mother gorilla caring for her baby, and a gorilla group making their sleeping nests. The final sequence was the capture of the two young gorillas that were sold to the San Diego Zoo. Osa Johnson cranked the camera while Martin and the porters cut down the tree in which the two young gorillas clung to each other, then threw tarps over the babies' heads as the tree crashed to the ground.

Dr. George Schaller (below left), in recalling his trek into the Virungas, relates, "Not being tourists, we were automatically classed as missionaries. Since one of the French words for "expedition" is "mission," I am sure that, with our far-from-fluent French, we sometimes conveyed the impression that we had come to convert the gorillas."
Dr. Dian Fossey (below right) was a 31-year-old occupational therapist when, in 1963, she decided to fulfill her lifelong dream of a safari to Africa.

(Top) Looming over the Manhattan skyline, King Kong dangles Fay Wray in a manner that looks most uncomfortable. Though King Kong was undeniably very big compared to his human counterparts, the story line itself was very much on the order of Beauty and the Beast. The terrifying creature was obviously enchanted by the lovely Miss Wray. Though he certainly scared her half to death, he never made the slightest move to harm her. Alas, his Beauty did not come to love him, and the Beast died, as in the end, we knew he must.
(Bottom) Extremely advanced film-making techniques were used to create the jungle battle scenes in which King Kong does hand-to-hand combat with prehistoric monsters like this Hollywood-version of the carnivorous dinosaur Tyranosaurus Rex. The dinosaurs, of course, never had a chance. Neither did reality, since no giant apes lived during the age of the dinosaurs.

The film was a great hit, but its depiction of the gentle giant apes was obscured entirely in the public mind by the raging success of another gorilla film. *King Kong*, never intended to be anything but fantasy, colored indelibly the public impression of gorillas for nearly fifty years. (In my hometown in Southern California, an enormous stuffed ape, easily ten feet tall, stood outside a huge second-hand store for years. My parents solemnly assured me it was the very ape used to make *King Kong*. It stood outside in all weathers, becoming distinctly threadbare with time. Having once conquered my fear, I came to feel enormous affection for the huge creature that served to announce that the store where I could buy books for a nickel was open.) What the vast majority of viewers remembered about King Kong was the gigantic ape that carried Fay Wray up the Empire State Building, only to be shot down by brave gunners in biplanes. A sillier movie has probably never been made, nor one with greater impact on the public notion of gorillas.

In an age when the daily headlines are more startling than anything that famous producer David O. Slelznick dreamed for **King Kong**, it is hard to imagine the film as having ever been promoted as a horror film. Funny, yes. Scary, no. In fact, the giant ape now seems more endearing than horrific, making posters like this one seem a nostalgic reminder of a more innocent age. Yet the age wasn't innocent at all; in 1933 the world was still struggling in the throes of the Great Depression, and the harbingers of World War II were already dimly visible on the horizon. For a few cents, for a few moments, **King Kong** took people away from the daily struggle of the present and obscured the fearsome future.

In the end, it took Dian Fossey's fourteen years of research on the mountain gorilla to turn public perception around. Her book, *Gorillas in the Mist,* an odd combination of solid scholarly research and personal anecdote, was a huge success when it hit the bookstores in 1983. Her brutal murder made animal conservation issues front-page news in every capital of the world, while speculation ensued over whether she'd been murdered by assassins hired by a corrupt government, by jealous researchers on her own staff, by poachers exasperated with her interference, or by persons unknown. The consensus finally came down to persons unknown, for her murder has never been solved. She died a terrible death, but her work may well save the greatest ape of all from extinction.

Actress Sigourney Weaver (far left) was nominated for an Academy Award for Best Actress for her role in **Gorillas in the Mist.** *Dian Fossey was murdered the day after Christmas, 1985. Buried December 31 in this grave at Karisoke (left), her eulogy was given by a missionary from Gisenyi, the Reverend Elton Wallace: "Dian Fossey was born to a home of comfort and privilege that she left by her own choice to live among a race faced with extinction. She will lie now with those with whom she lived, and among whom she died. And if you think that the distance that Christ had to come to take the likeness of Man is not so great as that from man to gorilla, then you don't know men. Or gorillas. Or God."*

On December 31, 1978, Digit, a gorilla Dian Fossey had known for ten years, was brutally murdered by poachers. Fossey had what remained of his body— the head and hands had been chopped off— buried in this grave at Karisoke (left). The murder led her to create the Digit Fund to finance ranger patrols to prevent further poaching in the park.

This lowland gorilla mother carries her three-month-old infant on her back in the "cool dude-Look, Mom, no hands" position.

Great Apes

It's easy to tell the great apes—orangutans, chimpanzees, and gorillas—apart at the zoo if you remember that they are color-coded. Orangutans are red, chimps are dark brown, and gorillas are black. Chimps and gorillas are both found in the rain forests of equatorial Africa, while orangutans are found on the other side of the world in Southeast Asia, specifically in Borneo, Sumatra, and Indonesia. In order of size, gorillas are the biggest of the great apes, chimpanzees, the smallest, and orangutans are in the middle. Gorillas and chimps are found in groups in the wild; orangutans roam through the trees alone. All the great apes are extremely intelligent: Chimps are clever; orangutans, contemplative; and gorillas, wise.

This lowland gorilla shows the distinctive nose print that Fossey used to tell the gorillas in the wild apart. Lowland gorillas were once found in central and western Africa, even on the coasts of west Africa, according to early reports.

(Far left) A Sumatran Orangutan mother eats a quiet meal while her fifteen-month-old baby clings to her back. As with gorillas, orangutan mothers are very protective of their babies. To collect the young ones preferred by zoos, it is necessary to kill the mother. (Near left) This adult male Sumatran Orangutan clearly shows the distinctive facial disk of the Orangutans.

*This pygmy chimpanzee (**Pan paniscus**) mother and child (far left), live in the forests of Central Africa, particularly in the Ituri forest. Chimpanzees (near left) are the most studied and best known of the primates. Because they are intelligent, extroverted, and easy to work with, most laboratory primate studies have been conducted using chimpanzees.*

MYTH,
LEGEND,
AND
HISTORY

C H A P T E R

The
Lowland
Gorilla

This portrait of a silverback lowland gorilla (far left) shows the dark, shiny skin on the face and the thick hair on the head that extends down the back. Like people, gorillas have distinctive faces, and facial resemblences are passed on from generation to generation. Gorillas walk mostly on all fours as this one is doing (left). Their hands are folded back so that they walk on their knuckles. Gorillas do sometimes walk on their hind legs, but that position is assumed either to gain height, to see better, like a person standing on tiptoe, or as a threat posture to warn off intruders. The threat posture is usually accompanied by loud roars and chestbeating.

D esmond Morris, former curator of mammals at the London Zoo, once wrote that farmers and hunters often know more about the life history and behavior of wild animals than zoologists. Morris just happens to have taken his doctorate in zoology, which means he's one of the few people who can get away with a statement like that. Today, long-term field studies observing the behavior of animals in their natural habitat are common, but in the fifties when Morris was doing his graduate work at Oxford, zoologists were either, in his words, "charming natural historians who pottered about in woods and fields, gazing rather ineffectually at Mother Nature, or white-coated laboratory psychologists so isolated from reality that they wouldn't know a field if they saw one."

The lowland gorilla is a case in point. Africans hunt them for meat, explorers have reported accidentally stumbling across them, big game hunters have shot them, and animal collectors have brought them back alive for zoos and circuses, but scientists have only rarely studied their life in the wild.

(Right) This lowland gorilla was photographed in Cameroon. Many of the gorillas in zoos and circuses around the world today were captured in Cameroon, including Michael, the male that was purchased to be a companion and mate for Koko, the gorilla that has learned sign language.

(Below) Two gorillas quietly enjoying a snack in Zaire. Gorilla infants are surrounded by constant security, care and protection.

(Far right) Disgruntled, I'd say. Seriously unhappy. Resistant, possibly even defensive. Do you suppose this was the shot that preceded the one on the next page? Or am I taking all those books on body language too seriously?

There are a number of reasons for this. The heart of Africa, where both species of the lowland gorilla lives, is an area dangerous to human health. "This section of the Dark Continent is one of the most pestilential regions of the globe. The very ground exhales disease, and sea captains have told me of the sickening odor of the malarious swamps [that] can be noticed while scores of miles from land," declared P. T. Barnum. Local medicine has historically been, and to a large extent still is, primitive in the extreme. The only medical help available in Barnum's day came out of the ill-furnished medical kits of local missionaries. There were, and still are, hundreds of poisonous snakes and spiders for whose toxins antivenins do not exist even today. Scrapes and scratches are slow to heal in this part of the world, and they often become infected. The heat and humidity are horribly oppressive. Even today, hacking one's way through dense, steaming jungle or poling dugouts along crocodile-infested rivers are almost the only means of transportation. The fearsome reputation of the gorilla has not exactly been an added inducement to spend extended periods of time in their company.

GORILLAS

One Victorian gentleman, eager to study the lowland gorilla and live to tell the tale, constructed a sturdy iron cage for himself, took it to the jungle, and climbed in day after day, patiently waiting for gorillas to wander by. None did. That was in 1896. Sixty years later Fred Merfield, a guide for big game hunters, published his tales of stalking the wild gorilla in *The Gorilla Hunter,* which was the best account to appear since du Chaillu's. In the ensuing years scientists mindful of their health have studied lowland gorillas in zoos in countries where the climate is better. There is a widespread rumor that some Japanese field studies on the western lowland gorilla are currently underway, but if so, the findings are as yet unpublished.

Big, bright, brown eyes expressing an alert, intelligent interest characterize this young lowland gorilla from West Africa as he hangs on tight to his mother.

Having read this far, have you begun to wonder how gorillas ever got their reputation for ferocity? This is how. He may be yawning or roaring, but most folks don't hang around long enough to find out. By the way, only males have the enlarged canines present here in all their glory.

Lowland gorillas stand between five-and-one-half and six-and-one-half feet (1½ and 2 m) tall—when they stand on two legs, which isn't often. Most of the time they walk on all fours, their hands curled back so that they can balance on their knuckles. In their normal stance, they are about four-and-one-half feet (1½ m) tall. Full-grown males (fifteen years old) weigh roughly 375 pounds (170 kg), and adult females about 200 pounds (90 kg). Young males (eight to thirteen years old) weigh about 250 pounds (114 kg), and young females about 170 pounds (77 kg). From the time they are babies until three years old, young gorillas weigh between two and thirty pounds (1 to 15 kg). Western lowland gorillas have black skin and thin, short hair on their bodies. There is no hair

on their faces, chests, hands, or feet. Zoologist Ivan Sanderson, who collected gorillas in West Africa in the mid-thirties, reported hair color that varied from black to silvery gray, and even saw some gorillas with bright red crests on their heads. Mature males have a silvery mantle down their backs and are called *silverbacks*; immature males are called *blackbacks*. The armspan of an adult male gorilla collected in the Virungas for the British Museum of Natural History was nine feet, two inches (3 m). The head was thirteen inches long (33 cm), and the ear two inches (5 cm). From the top of the gorilla's head to the base of the spine measured just over four feet (1.2 m), and the legs measured a little more than two feet (64 cm), for a total length of something over six feet (1.8 m).

This adult lowland gorilla (far left), photographed in Cameroon, takes a protective stance as a warning that it is prepared to defend the young one sitting next to it. Many animals, including gorillas, use size as a threat, by moving from a relaxed, sitting position to an attentive, erect posture thus making themselves look larger.
Gorillas forage for about two hours in the morning and two hours in the afternoon. Eating is invariably followed by resting, as this gorilla (left) demonstrates. Postprandial naps are not uncommon.

In skinning the animal, Sanderson reports that when the meat was completely cleaned from the bones, it made a pile as high as his waist. He also mentions that the worst of the skinning job was the hands and feet "because the tough skin on the palms and soles is bound rigidly to the bone and flesh beneath by a maze of ligaments as strong as wire."

Twenty men worked thirty-six hours to skin and clean the gorilla's skeleton; then the skin was stretched on a frame and dried over a slow fire for three days. The meat was divided among the people of the local village for food (the chief selected for himself a rump steak and the intestines), and the hunter took those portions necessary for his *ju-ju*, a tribal fetish, or magical rite.

Sanderson's description of the hunter's *ju-ju* is a fascinating glimpse into another culture:

> His knowledge of anatomy was remarkable; with amazing deftness he selected pieces of the muscular covering of the eye-ridges, flesh from the groin and armpits, the whole heart, a part of the small intestine, the tip of the left lung, some abdominal muscles, and the pancreas. I tried to fool him by offering the left lobe of the liver for the pancreas, which was exactly the same in color, but he was not having any of it, and after glaring at me as though I were only fit for a ritual *ju-ju* murder, he carried the pancreas off.

The purpose of the *ju-ju* is to make the hunter's aim more accurate. To this end, the hunter buys a recipe of various herbs, from which he makes a medicine that he keeps in a pot, surrounded by a small fence, within his compound. Some of this is rubbed on the man's gun before he goes hunting to help him find game and kill it. When he is successful, he takes certain parts of the animal he has shot, cooks them with a chicken, puts half in the pot with his *ju-ju* medicine, and eats the rest. With each successful hunt the man's *ju-ju* grows, and his aim becomes better.

The gorillas of Cameroon were once found along the coast, according to the reports of Carthaginian seafarers. An account found in the **Codex Heidelbergensis,** *written by Hanno in 525 B.C., may be the first report of gorillas. "On the island on the Southern Horn wild men appeared, among them many women with shaggy heads, whom our interpreters called gorillas."*

Of the estimated 14,000 lowland gorillas in the world, only 4,000 live in reserves or sanctuaries where they are protected. The countries in which they occur—Nigeria, Cameroon, Gabon, Congo, the Central African Republic, and Zaire—are among the poorest in the world, with rapidly increasing human populations. While some of these countries have passed laws forbidding the hunting of gorillas, enforcing the law in remote jungles where the people kill gorillas for food and for *ju-ju* is difficult, if not impossible.

Sanderson wrote:

> I shall never quite forget the emotions that this sight [of the dead gorilla] conjured up inside me. I had always been taught to think of the gorilla as the very essence of savagery and terror, and now there lay this hoary old vegetarian, his immense arms folded over his great pot belly, all the fire gone from his wrinkled black face, his soft brown eyes wide open beneath their long straight lashes and filled with infinite sorrow. Into his whole demeanor I could not help but read the whole tragedy of his race, driven from the plains up into the mountains countless centuries ago by more active ape-like creatures—perhaps even our own forebears; chevied hither and thither by the ever-encroaching hordes of shouting hairless little men, his feeding grounds restricted by farms and paths and native huntsmen.

Zoos in a number of countries have signed agreements not to purchase gorillas captured in the wild, participating instead in the Species Survival Program (SSP). The SSP is a captive breeding program linked to the International Gorilla Studbook at the Frankfurt Zoo, a studbook precisely like those for thoroughbred horses or pedigreed dogs. It lists all the gorillas in all the participating zoos around the world with as much of their breeding history as is known. In fact, all that is known of gorillas captured in the wild is their sex and where they were captured at what age—if even that much information is available. The zoos then exchange animals in order to create breeding pairs in the hope of breeding enough gorillas in captivity to supply the zoos that wish to exhibit them. The purpose is to eliminate collecting gorillas from the wild where so many are inevitably killed while defending their families from the collectors.

Andrew Battell, an English freebooter, told a story of apes like men, published in 1613. "They hurt not those which they surprise unawares, except they look on them. Their heighth is like a man's, but their bigness twice as great. He differs not from a man, but in his legs, for they have no calf. They feed upon fruit that they find in the woods and upon nuts, for they eat no kind of flesh."

The Species Survival Program is elaborately designed to maintain the greatest possible gene pool. Steps are taken to prevent close interbreeding, but still it is a tricky business. Females are particularly prized for their ability to produce babies. At the first sign that a baby is not thriving under its mother's care, the infant is "pulled," zoo jargon for taking the baby away from the mother to be hand-raised by humans. There are two problems inherent in this procedure. One is that hand-raised infants imprint on people, believing that humans are their parents, with the result that some females (like Pogo in the San Francisco Zoo and Koko, the gorilla who speaks in sign language) prefer people to gorillas when they are old enough to mate. Both Pogo and Koko have steadfastly rejected the advances of male gorillas (Ugh! He's such an *animal!*), but will flirt invitingly with people. (Much more about the fascinating Koko to come in chapter five.)

The other problem is that mothering is a learned behavior for gorillas, as it is for people. Wild gorillas live in close family groups where there are likely to be little ones of various ages. Females observe how to clean, nurse, groom, hold, and carry infants by watching older mothers. If a gorilla is not reared in a family group where she can see these things done, she never learns how, and cannot properly care for her infants, which must then be pulled, thus perpetuating the cycle.

These, of course, are only the physical issues of caring for a baby gorilla; there are also emotional consequences that must be considered. When mothers and babies are separated, it is traumatic for both. Susan Green's hundreds of pages of observational notes written when Patty Cake at the Central Park Zoo was taken from Lulu, her mother, made it unmistakably clear how painfully Patty Cake and her mother suffered from being separated. Lulu, heavily drugged, screamed and fought helplessly to prevent the zookeepers from taking Patty Cake away. Her mate, Kongo, in another cage went wild when he heard Lulu's screams. He tried desperately to open the doors between the cages to come to her. He cried, and ran to the windows to be able to see her. He banged on the windows and tore at the locked doors, and when that didn't work, he succumbed to diarrhea. (Diarrhea is common among threatened, terrified gorillas in the wild.) When Lulu recovered from the drugging, she searched the cage for her baby for days on end. She became listless and would scarcely eat. Patty Cake, too, showed signs of depression.

Gorillas feel and remember. Lulu and Patty Cake recognized each other when they were finally reunited three months later. Michael, the male gorilla purchased to mate with Koko, has related in Ameslan (American Sign Language) how he was captured when his mother was hit in the back of the neck by hunters, and the blood he saw running down her body.

There are other problems as well. Gorillas raised in family groups routinely see other gorillas mating. Gorillas kept in individual cages or hand-raised by people do not. In an effort to get their breeding-age gorillas to mate, the San Francisco Zoo some years ago resorted to showing the apes pornographic movies in the desperate hope that the gorillas would get the idea.

Gorillas in the wild do occur in what scientists call "family groups," but they are rarely the nuclear family structure of mother, father, and child as we see here at the San Diego Zoo. A more typical "family group" in the wild consists of one or two breeding males, a few breeding females, sub-adults of both sexes, and children from infants to toddlers.

THE

LOWLAND

GORILLA

43

Recently it has become popular in conservation circles to propose taking endangered animals from the wild to breed in captivity in the hope of preventing extinction, as has been done with the California condor. Dian Fossey refuted this theory emphatically by pointing out that three times more gorillas have been taken from the wild than have been born in captivity—and her figures did not include those members of the gorilla groups killed in attempts to take the young ones alive. Fossey advocated instead the establishment of game sanctuaries, parks and reserves, and strict enforcement of laws protecting the gorillas.

Unfortunately, the human population explosion in African countries has created the need for more land for people's houses, and more land to grow food. More land for more people means less land for wildlife of all kinds, and the loss of a place to go and a place to live that offers the kind of diet the animals eat is the greatest threat to gorillas. Conservationists call the problem habitat encroachment.

George Schaller surmised that western lowland gorillas in the rain forest of the Congo basin move about in adaptation to changing conditions:

> The most important disturbance to the region in the past 200 or more years has been the repeated clearing of small patches of the forest for cultivation. Africans work the fields for three or four years and then abandon them for at least ten years, allowing the forest to regenerate. Gorillas favor the dense secondary growth of these fallow fields, where forage in the form of herbs, shrubs, and vines is plentiful, over primary forest, which supports only a sparse ground cover in its shadowy interior. Many gorilla concentrations occur near roads and around villages where disturbance to the forest has been most recent.

If Schaller's observations are accurate, and there is every reason to believe they are, the lowland gorilla may have found a way to live in peaceful coexistence with the African need for subsistence agriculture. The two key issues in preserving the lowland gorilla are that the abandoned plots be left undisturbed long enough for the gorillas to find cover and forage in the secondary growth, and that the gorillas be protected from hunting.

CHAPTER

The Mountain Gorilla

Flossie (far left) was a mature female in Dian Fossey's Group Five. Dian named her after a favorite aunt who reportedly considered it a dubious honor. (This page) Gorilla children are cherished and indulged. There are many recorded instances of the group slowing down to accomodate a youngster's shorter legs, and of males adopting and caring for babies when the mother dies. Here, a youngster climbs all over a silverback with absolute impunity.

Mountain gorillas chuckle and smile. They purr with contentment, their great bellies vibrating like a cat's. They are intelligent and dignified. They feel deeply and remember for years. On the rare occasions when they quarrel, the disagreement lasts but a few moments. Even in serious confrontations, mountain gorillas fight to establish a winner and a loser, never to kill. Any fight may be ended at once by one mountain gorilla accepting the other's dominance. They are huge and powerful and endlessly gentle and patient. The most mischievous little one may scramble all over the most dignified silverback, poking him with knees and elbows, pulling his hair, and keeping him from a quiet nap, all with perfect impunity. Babies are the center of attention for the whole group from the moment they are born until they are about three years old and able to take care of themselves. Mountain gorillas care for their babies with great love and affection, dandling them, tickling them into giggling fits, and cuddling them often. They defend them with their lives.

When a group is threatened, the leader, usually a huge male silverback, rears up on his hind legs to stretch to his most imposing height, beats his chest, waves his arms, tears branches from trees and waves them, and roars. If the threatener, whatever it is, does not immediately break eye contact, drop into a submissive posture, or vanish, the silverback charges, stopping just short of the enemy. When the sight of a roaring, charging, arm-waving six-and-one-half foot (200 cm), 375-pound (170 kg) mountain gorilla is enough to cause the enemy to run for dear life, the mountain gorilla may give chase, grab the hapless creature, give it one or two good bites, and dash back to follow its group, which the distraction of the silverback's charging defense has enabled to flee screaming into the forest. (According to Dian Fossey, good gorilla etiquette is to hold your ground,

while looking as submissive and innocuous as possible; pretending to eat leaves is good.) The dominant silverback may be backed up by other silverbacks in the group, or by younger males, but ordinarily, these males stay with the females and little ones as they run to safety.

When a gorilla group is frightened, the males give off a very strong odor from their sweat glands that smells a lot like garlic. The fleeing gorillas may also exhibit a very liquid diarrhea. Combining these qualities with a charge complete with roaring, chest beating, and branch waving, it is easy to see why gorillas are rarely pursued by enemies other than people. Older references report that leopards prey on young gorillas, but neither Schaller, in his 18-month study, nor Fossey, in her 14-year study, observed any evidence of this.

Pablo (far left) may look the dignified silverback here, but Fossey ruefully recorded his impish nature when he was younger. Pablo once snatched her field notebook and mischieviously proceeded to eat it page by page. She searched the dung in his night nest the next day, hoping to retrieve something of her work, but to no avail.

(Left) As a youngster, Ziz's right hand was caught in a wire noose snare. Fossey speculated that his father, Beethoven, managed to free him, though the noose left cuts and scrapes that kept him from using the arm for a week.

There are twenty-nine slight physical differences between lowland gorillas and mountain gorillas. Dian Fossey points out that the differences are the result of adaptation to the higher altitudes at which the mountain gorillas live. The mountain gorilla has, for instance, longer body hair, a higher forehead, a longer palate, larger nostrils, a broader chest, shorter arms, and shorter, wider hands and feet than the lowland mountain gorilla.

Mountain gorillas live on the ground. They climb trees mainly to escape enemies (primarily people), occasionally to build sleeping nests, and to reach fruit. At higher elevations trees generally are not strong enough to support the weight of 200- to 400-pound (90- to 180-kg) mountain gorillas, so youngsters may do the fruit picking for the group. Mountain gorillas are vegetarians, eating leaves, bark, flowers, and the pith of stalks and roots, augmented by seasonal fruit and bamboo shoots. George Schaller observed gorillas eating one hundred different species of plants, while Fossey observed fifty-eight. Leaves, shoots, and stems make up 86 percent of the gorilla's diet, and fruit only 2 percent. The bulk of the mountain gorilla's diet is *Galium,* a thin, scraggly vine related to what we call bedstraw. The gorilla also eats a great deal of the bitter wild celery, prickly thistles, and stinging nettles. Schaller was wistfully envious of their ability to touch the nettles without getting nasty red welts. Gorillas also eat dirt to obtain necessary minerals such as sodium and potassium, and they eat their own dung, perhaps to obtain vitamin B_{12}, which may pass through the upper intestine without being absorbed the first time around.

Neither Schaller nor Fossey saw gorillas drinking water, but two American scientists, Frances Reynolds and Vernon Reynolds, reporting on the chimpanzees of the Budongo Forest, mention seeing mountain gorillas drinking in a film titled *Lords of the Forest.* Their drinking method is quite unusual, and common to chimpanzees, orangutans, and gibbons as well as the gorilla. "Each time the method was the same," wrote Reynolds and Reynolds. "Sitting facing the bowl (in a tree filled with rainwater), he repeatedly put his right hand into the bowl, pulled it out, and holding it in front of him, licked and sucked the water off his fingers."

Young mountain gorillas like Ndatwa (left) are more likely to climb trees than their elders, simply because most of the trees won't support the weight of an adult.
Stinging nettles (Urtica massaica) (right) are found most frequently at the western base of Visoke in a zone from one- to two-fifths of a mile (220 to 640 m) wide, as well as on the lower slopes. It is a favorite gorilla food.

(This page) *The role of the silverback—this is Ziz—is essentially that of a benevolent dictator. The dominant silverback has first choice of breeding females, decides where the group will feed, selects the night's sleeping site, and defends the group by fighting a rearguard action against intruders. Sometimes the alpha male will be backed up by other silverbacks or blackbacks to draw the enemy's fire as the rest of the group flees. (Opposite page) Here it is, the ultimate reversal: An ape (Joli Ami) is playing Tarzan.*

Unlike chimpanzees, gorillas do not hoot and holler for hours on end. Except when threatened, they speak softly or not at all. Fossey classified their sounds into roars, screams, cries, grunts, belches, hoots, chuckles, question barks, and *wraagh* sounds.

Mountain gorilla groups range in size from four to nearly thirty gorillas of all ages, from infants to the elderly. Each group is led by a silverback, usually the largest or strongest of the group. Dominance among males appears to be largely determined by size and strength. In groups with several silverbacks, the largest of them will lead the group. The silverback decides when the group will move, when it will stop, where it will rest, and where it will sleep for the night. He sets the pace of travel, and often slows everyone down so that a young, old, sick, or injured member can keep up with the group. Females establish their own hierarchy, but mothers with babies seem to be at the top of the pecking order with both older and younger females somewhere below.

Mountain gorilla groups are cohesive, staying together for months at a time. The groups Fossey saw were consistently smaller than those observed by Schaller. Some portion of the explanation for this lies in the fact that the mountain gorilla population was reduced by nearly 40 percent by the time Fossey ended her studies. Mountain gorillas have very distinctive faces, and Fossey identified the mountain gorillas in her study groups by their individual nose prints, sketching them in her notebook until she became familiar with each individual. Family resemblances help establish the relationships between the various gorillas in a group.

Some adult males strike out on their own, wandering alone. They may drop in on different groups and visit, staying a few hours or a few days, depending on how well they are accepted by each group's silverbacks. The groups' dominant silverbacks usually breed with the mature females, but not always. Breeding-age females that are consistently ignored by the silverback may hie themselves off to another group where their advances receive a warmer reception. Since female status in the group hierarchy is determined by having a baby, the transfer may allow the female to achieve a higher status within the new group. The fact that the silverbacks form the first line of defense when the group is in danger means that when they are facing guns or spears, they are often killed. As groups break up due to the death of the silverbacks, the females may transfer to other groups, or to a lone silverback to form a new group.

*Ndatwa munches contemplatively. Mountain gorillas eat fifty-eight different plants, of which eighty-six percent are leaves, shoots and stems, and only two percent fruit. **Galium** (bedstraw), a scraggly vine, forms the bulk of their diet, along with thistles, nettles, and wild celery.*

*Titus (above) invented chin-slapping, whacking his lower jaw with both hands to make a **clackety-clack** sound. One of the other youngsters joined in with hand-clapping. Reports Fossey,*

"Together the two sounded like a mini-minstrel band. On sunny, relaxing days, their claps and slaps could prompt playful spinning pirouettes from Simba, Cleo, and little Kweli."

Peanuts (in the foreground below) is the gorilla who reached out to touch Fossey's hand, the first voluntary contact ever initiated by a wild mountain gorilla toward a human being. **Robert Campbell, a National Geographic** *photographer who was with Fossey that day, captured the moment on film. Mountain gorillas typically spend thirty percent of their time eating and thirty percent traveling or travel-feeding—what we'd call eating on the run or grabbing a little snack.*

On a typical mountain gorilla day, the animals get up from their sleeping nests between 6:00 and 8:00 A.M., and eat enthusiastically for about two hours. At about 10:00 A.M., Schaller says, "activity slows down as most gorillas snack, sunbathe, and sleep intermittently until about 2:00 P.M." While the adults sit quietly or grab a quick nap, the young gorillas play follow-the-leader, king-of-the-mountain, and tag. They somersault down slopes and chase each other, chuckling as they run. They mock fight, wrestling, grappling, and tumbling about, and pretend to bite without hurting each other. For the hours between 3:00 and 5:00 P.M., the gorillas walk, snacking along the way. At 6:00 P.M. darkness begins to fall, and the group builds its nests for the night. Mountain gorillas spend about 40 percent of their time taking their leisure, 30 percent eating, and the remaining 30 percent traveling. Schaller reported that gorillas move anywhere from 300 to 6,000 feet (91 to 1829 m) a day within a home range of 10 to 15 square miles (26 to 39 square kilometers); Fossey found that the gorillas moved about 400 feet (122 m) per day.

(Left) Fuddle with Joli Ami from Beetsme's group.

Sleeping nests are made of branches and vines, broken or bent to form a crude bowl. The nests are cupped in the middle with a rim all the way around and are built either in trees or on the ground. Leafy branches are used to form a thick cushion in the bottom. Only plants not used for eating are used in nesting. In the rainy season, choice nesting places are in sheltered tree hollows, since gorillas are not fond of cold rain. (The mountain gorillas found at lower elevations, where the tropical hardwood trees of the Congo basin rain forest reach 180 feet [55 m], spend more time in trees than do the mountain gorillas of the high slopes of the Virungas, where the forest is mostly *Hagenia* and *Hypericum* trees that are substantially weaker and smaller.) Sanderson says, "The size of some of the platforms was quite staggering, and bore testimony to the great strength of these creatures." Sanderson says he found more than two dozen complete knots made in the creepers and saplings to keep them down, but neither Schaller nor Fossey found nests with knots.

(Above) In this photo, Shinda sprawls comfortably. Both day and night nests are constructed of plants the gorillas don't eat, particularly bulky ones like the giant lobelia (Lobelia giberroa).

Each nest is used once, since gorillas move on to a new nesting site each night. Gorillas defecate into their nests at night, a habit Osa Johnson declared "unspeakably filthy." Fossey, however, found the dung extremely useful in determining the age of a nest's occupant and also a reliable way to determine if births had occurred, since babies share nests with their mothers until they are about three years old, when the next infant is born. Babies begin making practice nests before they are a year old, and the youngest gorilla Fossey observed building a solid, serviceable nest in which he slept by himself was thirty-four months old. Mountain gorillas prefer to sleep on knolls and open slopes where anyone who approaches must hike up to the nesting site. The silverbacks do sentry duty, sleeping well below the rest of the group, making it impossible to approach the group undetected.

Here's a good look at a gorilla nest—this one happens to be Titus'. One researcher described the nests as looking like "leafy bathtubs." Stems are bent to form a rim, and leafy material is used to make a cushion in the middle. Remnants of tree nests last as long as four years, while evidence of ground nests lasts about five months.

Fossey figured that mountain gorillas rest forty percent of the day—and that doesn't include sleeping at night. Considering the terrain, it's no wonder. A gorilla's progress is uphill and down, day in and day out, and they don't even have pangas to hack their way through the undergrowth.

Fuddle (right) was absorbed into one of Fossey's gorilla study groups (Nunkie's) as an adult female. No gorilla group will endure long without a silverback to lead it. Loss of the silverback causes the group to split up and join other groups. Lone silverbacks (below) will seek to acquire females from existing groups in order to create a group of their own.

The mountain gorilla was discovered in 1902, two-and-a-half centuries after the lowland gorilla; by 1960 there was reason to fear that the mountain gorilla would become extinct in the same century in which it had been discovered.

Schaller explored the mountain gorilla's range for six months, after which he remarked that three-quarters of the species is "not found in the mountains, as their common name implies, but in lowland or equatorial rain forest at an altitude of from 1,500 feet to 5,000 feet [450 to 1,500 m]." He estimated, after a year and a half of research, that the total mountain gorilla population amounted to one mountain gorilla per square mile throughout their range—some 8,000 square miles (12,882 square km)—for a total population of 8,000. No one has estimated the *entire population* since.

In the Virunga volcanos, Schaller raised his figure to 2.9 mountain gorillas per square mile, for a total of 450 mountain gorillas in this small section of their territory. That was in 1960. In 1983, Fossey reported a population of only 240 mountain gorillas in the Virungas, though by 1988 the numbers had increased slightly to 275, according to Craig Sholley, director of the Mountain Gorilla Project. The number of mountain gorillas in the Virungas had dropped 40 percent in less than 30 years.

Mountain gorillas are afraid of snakes, and baby gorillas, who will cheerfully chase practically anything that moves, do not chase or touch chameleons or even caterpillars. Koko, the gorilla trained to speak in sign language, is afraid of crocodiles and alligators, even though she was born in captivity and has never seen either one.

There is substantial evidence that, left to their own devices, gorillas migrate vertically, foraging up and down the slopes of the Virungas. As the Park has been reduced in size, the gorillas have been increasingly confined to the higher elevations where certain foods, such as bamboo, are less available, and where the weather is colder and wetter.

*(Below) Dassies, as Tree Hyrax (**Dendrohyrax arboreus***) are commonly called, are delightful furry little creatures that look a lot like Hobbits. They inhabit hollow Hagenia trunks.*

(Above) Mountain gorillas love to sunbathe, a difficult thing to manage at the upper elevations of the Virungas where the rainfall is seventy-two inches (183 cm) per year.

Mountain gorillas also appear to be afraid of water. They will cross streams only if they can do so without getting wet, for instance, by using fallen logs as bridges. Fossey reports that they dislike rain and hunch in groups, looking miserable, during the downpours that frequently drench the mountains.

This fact, plus the statements of both Schaller and Fossey about being constantly soaked to the skin after tramping through the forest with the gorillas, piqued the curiosity of an American researcher, Charles James. What, he wondered, were animals who loved to sunbathe and who did not seem well adapted to a cold, damp climate doing in the Virungas at elevations of more than 10,000 feet (3,048 m) where it is often damp, misty, rainy, and chilly? Why did it often seem that Schaller's observations were at odds with Fossey's when both were excellent observers?

The clues were everywhere. Schaller had observed more mountain gorillas between 1,500 and 5,000 feet (450 and 1,500 m) than above 10,000 feet (3,048 m). He reported that they preferred secondary growth to the sparse undergrowth of mature forest. He counted one hundred plant species in their diet while Fossey counted fifty-eight. Fossey observed that the gorillas were susceptible to respiratory diseases, particularly pneumonia. She pointed out that the boundaries of the Parc des Volcans had been reduced in 1959 and again in 1966, and that the Rwandan government was considering reducing the Park by another 40 percent. Each time the park was reduced, the boundaries were moved higher up the mountain in order to open the lower slopes to cultivation.

The pieces of the puzzle began to fall into place. Perhaps the mountain gorilla is *not* native to the mountains; perhaps it is *retreating* up the mountains as the arable lower elevations are occupied by an ever-increasing human population. The higher the gorillas go, the less their traditional diet is available. At least half the plants they consider edible do not grow in the cooler climate of the higher elevations. While Schaller speaks of the gorillas moving *up* the mountain to eat bamboo shoots each season, Fossey mentions the gorillas going *down* to the bamboo zone. The gorillas are miserable-looking in the drenching rains because they're cold, and being cold and wet causes them to burn enormous amounts of energy just to maintain body temperature. That does not leave adequate reserves to fight off respiratory infections, which may eventually develop into pneumonia.

The higher the gorillas go, the less secondary growth there is, and hence the less food there is. Numerous animal behavior studies have demonstrated that crowded conditions cause social disruption. This may explain why Schaller saw no fighting between groups of gorillas, and Fossey saw a great deal. (Of course, it may also be that Fossey saw more fighting because she observed the animals for fourteen years and Schaller only for a year and a half.) Schaller's map of home ranges shows considerably less overlap between groups than Fossey's. Schaller describes the gorilla's dung as "quite solid, not at all messy, and the hair of the apes is not soiled by contact." Fossey reports numerous instances of diarrhea, a recognized ape response to stress and agitation. While Fossey itemizes a number of adaptations to the cold, wet climate of the mountains, where night temperatures can

(Far left) Fossey identified the gorillas she studied by first sketching their nose prints, and later by photographs when they were sufficiently habituated to let her get close enough to use a camera. She counted all the members of a group, named them, and sought to establish their relationships to one another. This is Inshuti of Group Five. Titus (left), who Fossey described as a "mama's boy," is now a handsome silverback. He is the son of Flossie, the gorilla Fossey named after her favorite aunt. Eating, as Titus is here, takes tremendous manual dexterity for a gorilla. Many of their foods, like wild celery and blackberry, must have their stems carefully peeled before they can be eaten. Others are wadded up before they're consumed. Gorillas also eat certain grubs and larvae, stripping the bark off trees with both hands and teeth to expose these delicacies. Gorillas also have a certain gourmet flair, indulging in wild mushrooms whenever they get the chance. Because it is a rare and treasured treat, squabbles tend to break out when one is found.

drop below freezing, there is no mention of adaptations to serious cold such as a thick layer of fat to maintain body heat, a coat of underfur to retain surface warmth, or a coarse or oily top coat of fur that sheds water. All of these points suggest that the mountain gorilla is *not* endemic to the mountains, as has been assumed, but instead migrates vertically following seasonal forage. With the establishment of the gorilla conservation area and the constant pushing back of the boundaries to higher and higher elevations, it may be that the gorillas have become trapped in an unsuitable environment that would prohibit an increase in their numbers even if all poaching and the capturing of wild gorillas were stopped entirely.

There is danger in this sort of speculation, but the fact remains that a close study of Schaller's and Fossey's work makes one wonder if they were studying the same animals. The differences between theory and practice are all too apparent in the case of the mountain gorilla. A substantial cause of the ever-shrinking park boundaries is the Common Market's decision to purchase pyrethrum (a daisy that produces a natural insecticide) from Rwanda in order to provide hard currency for that desperately poor country. The land to grow the pyrethrum has been taken from the Parc des Volcans, the sanctuary set aside for the gorillas, where Schaller did the first census of the mountain gorilla in 1960. The Rwandan sector of the Parc des Volcans sanctuary has been halved, reducing the Parc des Volcans by 20 percent. (The Parc des Volcans is made up of land in Zaire and Uganda, as well as Rwanda, and has a different name in each country.) Fossey declared that this loss of land and forage alone (without taking poaching into account) was enough to account for the 60 percent reduction in the mountain gorilla population in the Rwandan sector of the Parc des Volcans that she noted after Schaller's census.

(Left) Gorillas also use their teeth to peel stems. When the gorillas eat bananas found on abandoned farms, they eat not the fruit but the pith of the stem tearing the whole tree apart to do it. (Right) Wildflowers, like this one, are found in the rain forest and in the open meadows of the plateaus between the volcanoes.

George Schaller noted the ironic interaction between humans and gorillas: "Gorillas favor those parts of the forest which provide abundant forage on or near ground level. Under the canopy of the tall trees they find little to eat and consequently spend much of their time in the more open valleys and along rivers, where, in the sunshine, the undergrowth is dense. Especially suitable for gorillas are the abandoned fields near the villages, for there the apes find their favorite foods in greatest abundance. Man has here a curious role in the ecology of the gorilla. He is both an enemy in that he hunts the animal, and he is also inadvertently the ape's benefactor by providing it with optimum conditions for living."

The pyrethrum purchase is an example of one of those long-distance decisions, made in Europe to help Africa, with incalculably far-reaching consequences. Tiny Rwanda has 5 million inhabitants, the highest population density of any African country south of the Sahara. The population is increasing at a rate of nearly 4 percent each year, which means that every year 23,000 new Rwandan families need land for farms to feed themselves and grow pyrethrum as a cash crop. The pressure to terrace the mountains for more farms has two possible consequences, both appalling. The first is the destruction of the gorillas and all the other wildlife that currently lives there; the second is the destruction of the rain forest that provides the essential watershed. In Fossey's words:

> The real prospect of this destruction is as frightening as the fact that every year some 445,500 acres of the planet's rain forest are destroyed—a rate of 49.2 acres per minute.
>
> Foreigners cannot expect the average Ruandan [sic] living near the Parc des Volcans and raising pyrethrum for the equivalent of four cents a pound to look at the towering volcanoes, consider their majestic beauty, and express concern about an endangered animal species living in those misted mountains. Much as a European might see a mirage when stranded in a desert, a Ruandan sees rows upon rows of potatoes, beans, peas, corn, and tobacco in place of the massive *Hagenia* trees. He justifiably resents being refused access to parkland for the realization of his vision.
>
> American and European concepts of conservation, especially preservation of wildlife, are not relevant to African farmers already living above the carrying capacity of their land. Instead, local people need to be educated about the absolute necessity of maintaining the mountains as a water catchment area. The farmers need to know, not so much what foreigners think about gorillas, but rather that 10 percent of all rain that falls on Ruanda is caught by the Virungas and is slowly released to irrigate the crops below. Each farming family's personal survival depends upon the survival of the Parc des Volcans. If the importance of the ecosystem to the lives of the populace becomes a local priority, which is not now the case, the rain forest might stand a chance to survive, and with it, the animals it contains and the people who rely on it. As a country with much to gain both sociologically and economically from active preservation of land resources, Ruanda could well serve as a monumental example to Zaire and Uganda to gain fuller cooperation in safeguarding the Virungas that all three nations share.

Other solutions have been proposed, including increasing tourism to the gorillas to bring in hard currency. Zaire, which derives a substantial percentage of its hard currency from tourism, recently announced the decision to increase the park entry fees each year, the idea being to simultaneously find out what the market will bear, and bring in more dollars from fewer tourists. In the past few years the daily entrance fee to the Parc des Volcans has increased from $40 to $150, so go soon if you plan to visit. The increased fees are a way to protect the mountain gorillas; it does no good to tell impoverished people to cherish their resources if there is no benefit to them in doing so. If tourism brings in substantial amounts of hard currency to these countries, it will be in their governments' and their people's best interest to protect the gorillas, and this may be the most effective way of assuring the continued existence of mountain gorillas.

GORILLA
SAFARI

Located just south of the Equator on the high plateau of Central Africa, Rwanda is "blessed by a climate that's been compared to the French Riviera," according to the description in Overseas Adventure Travel's brochure. It is also blessed, if the term is not too sardonic, with one of the fastest growing populations in the world—more than five million people seek a subsistence living in a land barely the size of the state of Maryland. Every square mile (2.6 square km) must produce sufficient food and shelter for more than 400 people. Hills are terraced into two-and-a-half acre (1 hectare) plots with mud and grass huts called *shambas*. Once covered by thick forests most of those trees are gone today, cleared for planting or reduced to charcoal and firewood for fuel.

The Virunga Volcanoes are among the most
enchanting mountains in the world, rising
misty and snow-capped fourteen thousand feet
(4267 m) above Rwanda. The mountain gorillas
once roamed from the low hills up to the montane forests between ten thousand and thirteen
thousand feet (3048 and 3962 m) following the
seasons for foods that grew at different altitudes. In 1967, a European foreign aid project
to grow pyrethrum resulted in the size of the
Park being slashed by 24,700 acres (10,000
hectares), more than 35 percent, all at the
lower elevations. The idea was that Rwandans

could cultivate vegetables to sustain a family
on half the *shamba,* and on the other half, grow
pyrethrum (used to make a natural insecticide)
to sell to Europe in exchange for hard currency. To this end, the bamboo and hagenia
forests that had once sheltered the gorillas
were cleared for farms. The gorilla's natural
seasonal migration to lower elevations now
stops abruptly at the Park's boundaries, limit-
ing both their access to food and to the
warmer, dryer climate of the lower elevations.
It is ironic that the pyrethrum farming that was
to bring hard currency to this impoverished

land significantly endangers the mountain gorilla range despite the fact that tourism to see mountain gorillas is presently one of Rwanda's primary sources of hard currency.

The mountain gorilla safari is for neither the faint-of-heart nor the out-of-shape. It requires the ability to hike several miles uphill on a narrow, muddy, slippery trail carrying binoculars, cameras, lenses, film, raingear, gloves, dry socks, and lunch. Only three groups of six people each are permitted to enter the Park each day in order to keep disturbing the gorillas to a minimum. Once the group has climbed

to about 10,000 feet (3048 m), the true trekking begins. The terrain is very steep and very rough; slopes of 70 percent grade in and out of the ravines are not unheard of. Properly identified as montane forest, the description must be expanded to fit the facts of its geography. This is dense, often nearly impenetrable, Equatorial rain forest, as different from the open conifer montane forests of North America or Western Europe as the dunes of Death Valley from the beaches of Bali, though one could honestly say that both were sand. Deep ravines alternate with mountain ridges and grassy meadows.

The sub-alpine and alpine forest (above 11,000 feet or 3353 m) is mostly massive *Hagenia abyssinica* trees, thickly draped with long, dripping mosses and lichen, epiphytic orchids, and narrow-leaf ferns, until the form of the tree itself is scarcely distinguishable under what looks like a crazy beggar's streaming rags and tatters. George Schaller described Hagenia as looking like "a kindly unkempt old man." They grow to seventy feet (21 m) tall with trunks as much as eight feet (2.4 m) in diameter, have long, thin leaves and clusters of flowers the color of lilac. Other trees found in the Parc des Volcans are the *Senecto altioola, Senecia ericarosenii, Hypericum revolutum,* and *Vernonia adolfiiriderici.* At the lower elevations of the Park are stands of bamboo, which the gorillas eat when in season.

The forest understory is thick with vegetation often higher than a man's head. In many cases the only means of progress is to hack one's way through with a *panga* (machete). Most of the existing paths were once well-worn game trails pounded by the feet of thousands of elephants; as the elephants have been killed by poachers, the trails have grown over. Keen gardeners will recognize many of the understory plants as being cousins of those they know from their own gardens: *Clematis, Cineraria, Lobelia, Helichrysum* (strawflower), *Solanun* (potato/tomato family), *Thalictrum* (meadow rue) and *Rubus* (blackberry). The gorillas walk unscathed through large clumps of stinging nettle, apparently without even noticing, an ability that both Dr. Schaller and Dr. Fossey yearned for unreservedly. Fossey bemoaned the ability the nettles had of penetrating even layers of thick clothing, gloves, jeans, and heavy socks among them. She described other portions of the undergrowth as "brambles and thistles," and this is what today's wilderness adventurer must expect to trudge through to glimpse the gorillas.

Elevation takes its toll as well—it doesn't take much time or much exertion to find oneself gasping for breath at 13,000 feet (3962 m).

And then there's the climate. It rains. It rains a lot. And when it isn't rainy, it's drizzly or misty or foggy. The Virungas are undeniably lovely. And undeniably damp. The combination of high altitude and precipitation is an absolutely certain recipe for cold and wet. Both Schaller and Fossey reported routinely ending their days of gorilla tracking soaked to the skin and chilled to the bone.

The Virungas are essential watershed for Rwanda as well as the last remaining refuge of the mountain gorilla. Even if there were not a single gorilla or other form of wildlife there, it would be critically important to Rwanda to retain the rainforest watershed simply to provide water for daily use and for irrigating farmland. Streams, both year-round and seasonal, tumble down the steep ravines, across the grassy meadows, to find their way eventually to the terraced fields below. But issues of watershed, rainforest destruction, habitat encroachment, and wildlife preservation are issues of the industrialized nations of the First World. They are meaningless to the subsistence farmer, most often a woman, who is struggling to feed her family. Her children represent both her primary value in society and the closest thing to Social Security for her old age. She needs her children, especially the girls, to help her

work the farm. The boys are mainly hungry mouths to feed, but even given that they rarely earn their keep, the likelihood that she can be convinced to have fewer children is remote. Like morals, global issues are the luxuries of those who can afford them, not those who struggle daily for survival. The idea that land which might support a family should be preserved for mountain gorillas (an animal most Rwandans have never even seen) is welcomed about as enthusiastically as the suggestion that the buffalo (American bison) be restored to its natural range (from the hardwood forests of the East Coast across the prairies to the Far West) by removing the cities and farms now there would be by Americans. Yet the buffalo is a far more productive creature in terms of human use than the gorilla.

One possible solution to the problem is the development of eco-tourism. *If* those who can afford to travel to see the gorillas bring substantial amounts of hard currency into the country *and* that money reaches the people in a way that is significant to them, it then becomes in the people's interest to preserve the wildlife, the rainforest, and the watershed. If the money earned from tourism goes to the villages surrounding the perimeter of the Park in the form of schools, medical clinics, or cash grants to families, sufficient social pressure could be exerted on poachers to significantly reduce the threat to wildlife and wilderness.

W̲ould such a program fall to corruption? It's entirely possible. Billions upon billions of dollars in foreign aid and foreign investment have vanished in the last seventy-five years into the heart of Africa, absorbed like rain into the soil, and nearly invisible in its effect on village life.

There are many other kinds of wildlife besides the gorillas within the Park. Black-fronted duiker (*Cephalophus nigifronis*) and bushbuck (*Tragelaphus scriptus*) are small antelope that were the main game for poachers in Fossey's day. She describes at length the wire and hemp neck snares rigged to flexible bamboo stakes, the pitfall with sharpened stakes for catching buffalo, the booby traps of heavy logs that she hazarded a guess might be for bush pig. There were elephant and Cape buffalo as well, but the elephant's ivory, worth a year's wage or more to a successful poacher, condemned it to death in the Virungas as it has throughout Africa. There are few left now, though the Cape buffalo remain. And as the game in the lower elevations became scarce through hunting and clearing of the land for farms, the hunters followed its retreat up the mountain into the Park, becoming, by the act of crossing the Park boundary, poachers instead of hunters.

The gorillas, too, are victims of poachers, the silverbacks killed to make magic potions *(sumu)*. Whole family groups have been speared to death to capture a single baby for sale to unscrupulous zoos. On a smaller scale, there are tree dassies, the little *Hyrax* that is often described as the elephant's closest living relative, the clever white-collared ravens, and tiny brightly colored African frogs. There is a lot to watch for as one hikes, simultaneously keeping a wary eye out for cleverly concealed pits and snares.

This, then, is what the gorilla safari promises. Difficult hiking on slippery, muddy trails, across running streams and dry streambeds, through rugged, steep terrain, frequently wet, frequently cold, stung by nettles, scratched by brambles, pricked by thistles. And to top it all off, it's expensive. Is it worth it? That's a perfectly reasonable question to ask.

Y et, oddly enough, all reports agree: The
moment the tracker finds a fresh gorilla print,
all awareness of discomfort vanishes instantly.
The group is hushed and eager with anticipa-
tion. They are as anxious not to show anxiety
as they are to see gorillas. At first only a
glimpse of black pelt is all that's visible. Then
a patent-leather face with deep brown eyes can
be picked out of the foliage. The guide tells
everyone to bend over or squat down and
make soft, grunting noises, what Fossey called
"contentment vocalizations". As silly as it looks
and sounds, it feels even sillier, but no one hes-
itates for a second to do it. The gorillas watch
curiously. In time they begin to emerge from
behind the trees and bushes to look everyone
over. First come the young males, then the silverback, and last, the females with little ones.
The guide warns everyone not to stare; gorillas
interpret direct staring as a threat. Eventually
the gorillas begin to snack on wild celery.
The guide suggests we follow suit, in pantomime. It only takes one experimental bite to
find out why it's only a pantomime: Wild celery is amazingly bitter.

Suddenly a small gorilla, who has been eyeing the group, intently lets loose of its mother and trundles over for a closer look. Cameras and binoculars suddenly vanish into knapsacks as those curious little hands reach toward them. Other cameras, far enough away to be safe in their owner's hands, are snapping furiously. Deprived of the shiny things to investigate, the little one settles for boot laces, industriously tugging on them until they untie, and then tugging some more. Its mother watches the whole performance with remarkable calm and good nature. I certainly could not allow *my* toddler to poke at a bunch of gorillas with equal equanimity! When she decides the game has gone on long enough, she rises with a grunt, shambles over, and scoops the youngster onto her back. In the meantime, the owner of the boots has grown quite pale, but whether it's from pure thrill or pure terror is never clearly established.

On the long trek back to the lodge, every moment of the hour and a half observation is discussed, analyzed, examined from every possible angle, subjected to every possible interpretation, and then reviewed again. Perhaps everyone was cold and tired and wet and hungry when they got back. It seems likely. No one mentioned it at the time or recalls it clearly now. It wasn't important.

Visiting the Mountain Gorillas

To see the greatest of the great apes in the wild, contact Safari Consultants, (800) 544-6130, or 83 Gloucester Place, London, England W1H 3PG, telephone 935-8996, or Overseas Adventure Travel, 349 Broadway, Cambridge, Massachusetts 02139, telephone (800) 221-0814, about a safari to the mountain gorillas in Rwanda or Zaire.

CHAPTER 4

Captivating
Captives

photos: © Ron Garrison/Zoological Society of San Diego

(Far left) Trib, a male lowland gorilla, nibbles delicately on a handful of leaves at the San Diego Zoo in San Diego, California. Gorillas are strict vegetarians, eating only fruits, vegetables, and cereals. Zoos often supplement their diets with vitamins as well. (Left) Gorilla babies are hitch-hikers almost from birth until they are two or three years old. This big-eyed little one, photographed at the San Diego Zoo, thinks piggyback is the standard mode of transportation. We can only guess what its Mom thinks.

W hen the first reports of gorillas began filtering back to civilization, human curiosity took its natural course: People wanted to see a live one. Nineteenth-century zoos and circuses made every effort to obtain a gorilla to exhibit. Nineteenth-century zoos were menageries, collections of unrelated animals stuck in bare concrete-and-iron cages that people paid to stare at. Little was known about how the animals lived in the wild, and unless the information added a sensational interest that would draw customers, most zoos cared little. People who came to view the animals were no more enlightened. They bought peanuts, popcorn, and cotton candy and tossed them to the animals to eat. And the animals were teased mercilessly to drive them into a rage, which spectators found deliciously frightening from the safety provided by the bars of the cage.

(Right) The American explorer Martin Johnson, seen here with the pygmies of the Ituri Forest, was the first to show the world how unaggressive gorillas are in the wild. His movie, **Congorilla,** *was a great hit, but the impression it gave of the shy, gentle giants was rapidly superseded a few years later by* **King Kong.** *(Far right) Osa and Martin Johnson, and Belle Benchley, Director of the San Diego Zoo, visit one of the two young mountain gorillas the Martins sold to the San Diego Zoo. At the time, there were no other mountain gorillas in the United States; today, too, there are none.*

© Museum of Modern Art/Film Stills Archive

Zoos purchased wild animals from dealers, who bought them from local people in the animals' habitats or commissioned their capture. Many people were fascinated by these tales of adventure. Books like Frank Buck's *Bring 'Em Back Alive* and Osa Johnson's *I Married Adventure* became radio serials and ran for years.

Martin Johnson became convinced in the process of filming *Congorilla* (see Chapter 1) that the tales of the gorilla's ferocity were largely fabrication. He questioned Africans extensively about the legends that gorillas carried off women and killed men. The tribal Africans obviously thought these stories hopelessly silly, as told in *I Married Adventure.* "Why should gorillas carry off women?" , they asked. "Women are made only to carry firewood, plant gardens, and build houses, and gorillas don't have fires, houses or gardens." Martin was determined to take a gorilla back to the United States to "debunk all those stories of the viciousness of the gorilla."

When Osa and Martin Johnson captured a pair of baby mountain gorillas in the Belgian Congo and sold them to the San Diego Zoo for $15,000, they donated the money to build a new gorilla habitat. Though San Diego was far from being a major city, the San Diego Zoo was the newest of the major zoos in the United States, and Belle Benchley, the zoo's director, told the Johnsons about the facility's family group displays, spacious cages, year-round supply of fresh fruits and vegetables, and excellent record of good health and long life for its animals. These were revolutionary notions in zookeeping in 1931, and they were exactly what the Johnsons were looking for for the two little gorillas, Congo and 'Ngagi. When they left for their new home there were only three other gorillas in the United States, none of them mountain gorillas.

The year 1931 was big for gorillas. In that year Augusta Maria Daurer DeWust Hoyt went gorilla hunting with her husband in French Equatorial Africa to collect a big male gorilla for the American Museum of Natural History in New York. Bagging the silverback her husband wanted for the museum involved killing an entire gorilla group. The Africans hired to help speared the gorillas to eat; all but a little female, only a few months old, deemed too small for food. An African chief handed the struggling baby to Mrs. Hoyt, and it clung to her frantically. The Hoyts named her Toto, the Swahili word for "child", and raised her like a child until she was full grown.

The same year another baby gorilla, a male so sweet-tempered it was named Buddha, or "enlightened one," was brought back to Boston by Arthur Phillips, captain of the *West Key Bar*, a trading ship. Buddy, as the baby was called, survived the sea voyage from Africa to America beautifully, unlike so many young apes before him. Then tragedy struck. A sailor, sacked by the captain as they sailed into Boston Harbor, took his revenge by squirting a fire extinguisher of nitric acid at the little gorilla. The acid burned Buddy's face badly, damaging the facial muscles, and scalding his eyelids so terribly that he couldn't close his eyes. Gertrude Davies Lintz, a wealthy Bostonian who collected both chimps and gorillas, took the baby in and nursed it back to health. (Mrs. Lintz's other gorilla, Massa, was sold to the Philadelphia Zoo, where he became the longest-lived gorilla in captivity, dying at the ripe old age of 55.) Though it eventually healed, Buddy's face was permanently deformed into a ferocious grimace. That grimace was destined to become one of the most famous faces in America, for Buddy was sold to the Ringling Bros./Barnum & Bailey Circus, where a public relations man renamed him Gargantua the Great and put his snarling face on circus posters all over America. The public flocked to see "the world's most terrifying living creature," as the posters described the mild-mannered Buddy. He is credited with almost single-handedly snatching the circus back from the brink of bankruptcy in 1939 and 1940.

By 1941, the circus wanted a new novelty: a bride for Gargantua. Mrs. Hoyt, finding a full-grown gorilla difficult to manage, reluctantly agreed to sell Toto to the circus. She demanded that Toto be given the same luxurious air-conditioned accommodations that Gargantua enjoyed, and busied herself preparing an elaborate trousseau, including shirts, sweaters, socks, sheets, and blankets, all lovingly embroidered with *Totito*. Despite the elaborate publicity, no one dared put the two gorillas together in the same cage for fear they would hurt each other, and they lived side by side until Gargantua's death in 1949. Toto lived another twenty years, dying at the age of thirty-six in 1968.

Over the past fifty years zoos have undergone a radical change. They have moved from offering only entertainment to offering education about rare and unusual animals. They have shifted from collecting animals in the wild—and sometimes contributing to their endangerment in the process—to internationally coordinated breeding programs. Breeding exotic animals in captivity allows scientists to learn more about the animals and provides more animals for exhibit through exchanges with other zoos. The United Nations–sponsored Convention on International Trade in Endangered Species (CITES), signed by virtually all European nations and the United States, prevents wild animal dealers from capturing endangered species in the wild and selling them to zoos. There are flaws in the best-intentioned agreements, and the flaw in this one is Spain, one of the few European nations that did not sign CITES. Wild animal dealers sell endangered animals, especially the great apes, to Spanish zoos, which in turn sell or exchange them with zoos in other countries.

(Opposite page) Looking absolutely terrifying, Gargantua the Great was a huge draw for the Ringling Brothers/Barnum & Bailey Circus. In fact, "Gargantua the Great" was the brainchild of a public relations flack, the gorilla was actually called "Buddy", and was much-loved for his sweet temper and affectionate nature. The impressive snarl was the result of acid burns that deformed his face.

While there are currently no mountain gorillas in captivity in the United States, breeding of lowland gorillas is proceeding apace. The first gorilla born in captivity was Colo, born on Christmas Eve of 1956 at the Columbus, Ohio, zoo. The second, Goma, was born three years later at the zoological gardens in Basel, Switzerland. Both were females, and both were taken from their mothers to be raised by people. Two years later, in 1961, Goma's mother, Achilla, gave birth to a boy, Jambo, the first baby gorilla to be raised by its mother in captivity.

The ability to breed gorillas in captivity has been the source of virtually all the information available on the lowland gorilla, and it quite naturally has become a major focus of interest for those seriously involved with the greatest of the apes. Jambo was sold to Lawrence Durrell, the writer and naturalist. His zoo on Jersey in the Channel Islands of Great Britain was started with a loan backed by his publisher based on Durrell's reputation for hilariously funny, best-selling books about animals. Durrell has been among the most vocal supporters of captive breeding programs that reduce or eliminate the need for capturing wild animals, whether endangered or not, and his small zoo has been remarkably successful with gorillas.

Jambo came to Durrell in a typical breeding exchange, a process Durrell describes as being as delicate and complex as an arranged royal marriage. Durrell had two healthy females of breeding age, Nandi and N'Pongo, and a magnificent new gorilla habitat. All he needed was a healthy male of breeding age. Ernst Lang, director of the Basel, Switzerland, zoo proposed Jambo. Lang sent photographs, and described Jambo as "exceptionally powerful and exceedingly handsome, black but comely, and with a rather humorous expression." Best of all, Jambo was already familiar with the business of love, having successfully fathered a daughter at the Basel zoo.

(Below) Zoos have been completely transformed in the past seventy-five years. Animals are no longer imprisoned in small iron-barred, concrete-floored cages. Today most zoos strive to build naturalistic habitats like this one, the World of Primates at the Philadelphia Zoo. The emphasis has shifted from profit and entertainment to public education and species preservation through breeding programs.

Making introductions between gorillas is a delicate business, requiring all the arts of diplomacy, and, should those fail, hoses, buckets of water, and pitchforks. Standard operating procedure is to put the gorillas in adjacent cages where they can see each other, but not do each other any harm. The Jersey zoo's introduction came off well: Nandi fell instantly, hopelessly in love, and even N'Pongo, who generally thought Jambo was a nuisance (and with good reason: he teased her mercilessly), decided that he might have his uses when she came into season. Both Nandi and N'Pongo delivered healthy baby boys. In three years they delivered six healthy babies between them, a remarkable record for the Jersey zoo, considering that in all the zoos in the world there had been only seventy-four births in twenty-one years.

One of the key ingredients for producing a healthy baby—human or gorilla—is proper prenatal nutrition. Attempts to breed gorillas failed until research at the Philadelphia Zoo revealed that the captive primates' standard diet lacked essential vitamins, minerals, and trace elements. A primate vitamin pellet was devised, and breeding improved as did the animals' health and longevity. More work was done at the Basel zoo, resulting in the first successful captive gorilla breeding program in Europe and the birth of Jambo's older sister, Goma. Gorillas are notorious for disliking new foods, both in the wild and in captivity. Durrell tried introducing the Basel zoo's version of vitamins, a soft cookie an inch wide and a half-inch long at the Jersey zoo. The gorillas rejected it unanimously. "They displayed all the symptoms of appalled horror you would expect from a missionary to whom you had offered human flesh *en casserole*," Durrell reports. Gorillas in captivity are fed foods ranging from raw eggs and fresh fruits and vegetables to cookies and soda as treats, and they are no less fond of sweets than the average four-year-old child. The Jersey zoo added aniseed flavoring to its vitamins, which made the pellets taste a lot like licorice, and the deed was done.

(Above) The World of Primates at the Philadelphia Zoo has a large outdoor area with trees and grass for its gorilla group. Zoos are now making an effort to duplicate the social structure of animals as they are found in the wild. In the case of gorillas, that means family groups with a silverback, young adult males and females, and mature females who will produce babies.

CAPTIVATING
CAPTIVES

There is something infinitely appealing about gorillas. It is easy to see our kinship with these huge, dark creatures. Close observations made of gorillas in captivity have shown that it is a real cruelty to keep these highly social creatures isolated from one another. Until relatively recently little was known of the structure of gorilla groups in the wild, so, because they are expensive animals to obtain, they are often separated, even today, to prevent any chance of one gorilla seriously harming another. However, Desmond Morris points out that gorillas have a mild hierarchy system in the wild and that the silverback, the dominant male, functions as a benevolent dictator. Even in dominance battles between two males, gorillas never fight to the death. Combat is largely ritualized, an exchange of threat displays. With this understanding, the trend today is toward exhibiting gorillas in family groups similar in structure to those found in the wild.

The changes in theories about keeping wild animals in captivity are clearly demonstrated by this photograph of the gorilla enclosure at the San Diego Wild Animal Park, particularly when it is compared with the photograph on page 101. That photo was taken in the early 1930s, and the San Diego Zoo was considered radically progressive in its approach to zookeeping at the time.

Scientists and philosophers have spent countless hours trying to determine what, exactly, separates man from beasts. The biblical injunction that man shall have dominion over beasts has wreaked an ecological havoc even God could hardly have dreamed of. Secure and content in the belief of being but little lower than the angels, people have destroyed what God created with reckless abandon, rendering extinct hundreds of animal species. The problem comes when we speak of the "lower animals" as though there were a hierarchy, and those below us have no right to exist. It is only recently that people have come to glimpse the complexity of the interdependent relationships between ourselves and other living things, plant and animal alike. It is only recently that we have come to understand that we are merely contemporaries with other living things—living on the same planet at the same time, neither higher, nor better, nor more important.

Shango rides on Bawang's back while Pogo takes up the rear. While researching this book, I found no mention of babies riding on their Mom's backs this way, though I had observed Shango doing it at the San Francisco Zoo in San Francisco, California.

GORILLAS

A Visit to the Gorillas at the San Francisco Zoo

The San Francisco Zoo's Gorilla World has eight gorillas: Bwana, the silverback; Pogo, a mature female who was raised by missionaries in Cameroon and refuses to mate with gorillas; Mkubwa (Kubie), a blackback who is one of Bwana's sons; O. J., a young blackback; Bawang and Zura, both mature females; Binti Jua, a baby girl (Bwana's granddaughter and Kubie's niece); and Shango, a brand-new baby boy, born to Bawang and Bwana. O. J., Bawang, and Zura were all introduced to Bwana, Pogo, and Kubie in 1984. Now Binti, daughter of Bwana's son, Sunshine, is being introduced to the others through an exchange program with the Columbus, Ohio, zoo. She is fifteen months older than Shango. Binti was hand-raised while Shango is being raised by his mother, Bawang.

Not yet able to crawl, Shango scoots along on his belly to explore the rooms inside Gorilla World. Carried on Bawang's back, he affects a "cool dude" position never before reported either in the wild or in captivity: He stretches out on his back, arms folded casually under his head, knees bent and toes clasped. This projects a rather different attitude than the traditional methods of riding piggyback clutching his mother's fur or being carried held to his mother's chest.

It is poor gorilla etiquette to stare directly at a gorilla, but it was hard not to when I got the chance to watch the gorillas being fed dinner and bedding down for the night. Bawang brought her new baby close to the wire so we could get a good look at him. I could easily have touched him, had touching been allowed, but of course it wasn't. Bawang pushed her own fingers through the wire, reaching out to us, and when we did not reach back to her, she offered us presents: a delicious acacia branch first, and when that was returned by Carol Martinez, the senior primate keeper, she proffered one of the burlap bags with which the gorillas build night nests. Carol patiently pushed it back into the cage again and again. The gestures of friendship were so clear, the desire to touch so touching, that I felt obliged to apologize for not returning the gestures. "I can't. I'd love to, but they won't let me," I muttered over and over, holding my hands behind my back lest they betray me.

Binti's keeper brought her back in from her private playtime outside, a special time just for her, until she is fully integrated into the gorilla group. Binti didn't much like the idea of going back in the cage; she threw a screaming and kicking temper tantrum that would have done any "terrible two" proud. Zura came over to the wire between them and offered to play with Binti, but the baby kept her distance. People are still more familiar to her, and she rapidly climbed up the wire closest to us to see if we'd let her out and play with her. Alas! to no avail. Though we would cheerfully have picked her up and petted her as long as she liked, she'll never learn to live with other gorillas that way. I left, feeling that the human race had come off badly in the encounter. In the demonstration of intelligent interest and gracious friendliness, the gorillas had us beat by a mile.

GORILLAS

110

(Below) Bwana is the dominant silverback, the alpha male of the gorilla family at Gorilla World at the San Francisco Zoo. He is the proud papa of Shango, the baby boy on the opposite page. Bwana has another son, Kubie, in the group at Gorilla World.

(Above) Shango sleeps quietly on his mother's belly, as Bawang keeps a watchful eye on the photographer. Bawang was purchased from the Cincinnati Zoo in 1981, and Shango was born in 1989. Shango is being raised by Bawang in a family group that approximates gorilla groups in the wild as closely as possible.

© Steve Underwood Photography

112

Once the primates were recognized as evolutionary relatives of human beings, and the great apes as our closest kin, what then distinguished man from the apes? Some suggested that only people used tools. Then we found that Darwin's finches, certain anteaters, and chimpanzees all use tools. Next man was defined as the most intelligent of the animals, but that, too, fell by the wayside as experiments with whales, dolphins, elephants, and apes suggested an intelligence as great or greater than ours. This was a sticky one with gorillas. Not many scientists were eager to hop into a cage with a four-hundred-pound (180 kg) gorilla. And gorillas had a reputation for being all brawn and no brain. Chimps were easier to work with, extroverted and eager to please the human experimenters, and they were a more manageable size. Unlike chimpanzees, gorillas are massively indifferent to what people think of their intelligence. Jean-Pierre Hallet, a sociologist who spent twelve years in the African bush, described them in his book *Animal Kitabu*, as acting "like Buddhist priests forced to attend a football game." They are more concerned with accomplishing what they want to accomplish than what some laboratory psychologist wants them to accomplish. San Diego Zoo conducted tests with a fifteen-year-old silverback named Albert to determine whether he could distinguish between a circle and a square. The tests repeatedly ended with Albert smashing the testing equipment with his fists. Yet Belle Benchley tells a story in her book about her career as director of the San Diego Zoo, of 'Ngagi and another male gorilla that clearly demonstrates that gorillas are perfectly willing to use their intelligence to serve their own purposes.

> Our gorillas are apparently not interested in anything of a mechanical nature. When we put impediments in the way of their doing what they desire, they appear to lose that desire. At first this deceived us all, but one day when we had left a gorilla shut in for a particular purpose, the one that was left outside walked over to the door, which had confined them for a period of six years without being padlocked or held in place by bars, and, pressing his hollow palms flat against the door, pushed it to the top and held it there until the other gorilla had walked out. Then, withdrawing his hands quickly, he let the door drop and walked away with no further interest in his feat.

It is, perhaps, inaccurate to assume that because gorillas do not do as we wish they do not understand. There is always the possibility that they understand perfectly well and don't *want* to point to circles and squares for hours on end. This, in fact, may be a sign of greater intelligence. And it raises the whole issue of human arrogance in our assumption that we are even qualified to judge an ape's intelligence. Irven DeVore, the brilliant animal behaviorist, explained it this way:

> Suppose a New Yorker were to be trapped by a group of chimpanzees, shipped to Africa and stuck up in a tree a hundred feet above the ground. Practically all his abilities—his mastery of language, his skill at fixing a disabled fuel pump, his aggressive salesmanship—would be irrelevant to his situation. Hanging on for dear life, frequently confusing edible with poisonous plants, and, no doubt, experiencing grave difficulties distinguishing one chimpanzee from another, he would appear to his captors to be an exceedingly stupid animal. Their judgment, of course, would be unfair, since it would arise from a failure to appreciate that New Yorkers are not used to living in trees.

(Far left) Some gorilla babies must be taken from their mothers to be raised by people. It usually happens when the mother shows no interest in the infant, does not know how to care for it, handles it too roughly, or the baby becomes ill. The usual approach is to treat them like human babies. This is Gordy, complete with crib, toys, and diapers, in the nursery at the San Diego Zoo.

CHAPTER 5

Koko:
Bridging the
Gap Between
Man and
Ape

Taught American Sign Language by Dr. Francine (Penny) Patterson, Koko became the first gorilla to use language to communicate with people. Koko's pet kitten (seen on page 115), a birthday present, became a national celebrity with the publication of Patterson's book for children, Koko's Kitten. Huge as Koko was by comparison, she was always very gentle with the tiny kitten. When the cat was killed by a car, Koko asked for a new kitten for Christmas. (Far left top) Penny Patterson carries on a sign language conversation with Koko. (Far left bottom) Koko's vocabulary is constantly growing, and is now well over 500 signs. Here Koko asks for a drink by signing "sip". (Above) Koko looks at her biography. Koko is learning to read, though in this photograph she is probably just looking at the pictures.

S ince the ability to use tools and native intelligence did not adequately separate people from apes, the next distinction offered was language. The premise was that humans were the only animals to use language. Since neither gorillas nor chimps are believed to be able to make the sounds required to speak human language, comparative psycholinguists Beatrice and Allan Gardner decided to try to teach a chimpanzee named Washoe to use American Sign Language (Ameslan or ASL). The experiment, begun in 1966, was showing fascinating progress in 1971 when the Gardners lectured at Stanford University in Palo Alto, California. In the audience was a young graduate student, fresh from the University of Illinois, named Penny Patterson. As the Gardners spoke of Washoe's eagerness to learn sign language, Patterson felt she was hearing something straight out of the ancient myths in which the beasts of the world could speak. "Animals were capable of telling us about themselves if only one knew the proper way to ask them," she says.

That same year Patterson was invited by one of her professors at Stanford, Dr. Karl Pribam, to join him on a trip to the San Francisco Zoo to see about doing a computer-based language study with gorillas. While Dr. Pribam chatted with the zoo officials, Penny watched the gorillas. Bwana had sired a three-month-old baby girl, born July 4, 1971, named Hanabi-Ko, Japanese for "Fireworks Child." The baby, called Koko, was having a difficult day. Every time her mother put her on her back, the baby would slide off again and cling fiercely to her mother's belly. Over and over the baby slid off; over and over her mother pushed the baby back on her back, ignoring Koko's whimpers and protests. Patterson decided then and there to work with Koko and teach her sign language. Today Dr. Francine (Penny) Patterson is famous as the woman who taught a gorilla to speak.

Project Koko began in July 1972 when Koko was exactly a year old. American Sign Language is a series of hand gestures, each gesture signifying a single word. Patterson began by using sign language and speaking English simultaneously, and she asked the volunteers who were hand-raising Koko to do the same. Within only two weeks the volunteers reported that Koko was making signs. Patterson couldn't believe it happened so quickly and so spontaneously. In retrospect, she thinks that Koko really *was* trying to sign. Koko's formal education began a month later with Patterson molding Koko's hands into the shape of the ASL sign for *drink* while showing Koko her bottle. The first three words Patterson tried to teach her were *drink, food,* and *more.* It didn't take Koko long to catch on. Within two days she was making a fair approximation of the *food* sign while Patterson offered her tidbits of fruit. Patterson recalls that she was so delighted with Koko's progress that she showered Koko with praise and treats. Koko responded correctly through a long lesson, until by afternoon, the thoroughly stuffed little gorilla no longer had the faintest interest in food. Within two months Koko was using two-word combinations, such as *food drink* (her word for her formula, a combination of cereal and milk), and *food more.* It had taken Washoe the chimpanzee ten months to use two-word combinations.

Koko also began asking questions, using eye contact, cocking her head, and raising her eyebrows to signify a question. She loved to have Patterson blow a patch of mist onto the glass wall of the nursery at the San Francisco Zoo, so she could draw pictures in it. Her pictures were little more than squiggles, but she loved the game and learned to ask Patterson to blow on the window for her.

There are some limitations to sign language for gorillas, the greatest of which is that gorillas' hands are not shaped like people's. The thumb is shorter and placed farther down on the palm and farther away from the other fingers, so Koko's thumb won't reach her little finger. This hasn't presented much of a problem: Either Patterson has modified the sign so that Koko can form it with her hands, or Koko invents a sign that is similar. These modified signs make up Gorilla Sign Language, or GSL.

Koko habitually makes signs that involve motion, such as *long,* by starting the sign close to her body and moving her hand away, instead of extending her arm and drawing her hand toward her body. Autistic children often do the same thing. Koko also sometimes substitutes a similar sounding but inappropriate word when the right one does not come readily to mind, something autistic children taught sign language also do. The connection between gorillas and autism is an intriguing one. Dian Fossey worked for many years with autistic children before going to Africa, and she attributed her success in habituating the wild gorillas to her presence to her experience with the autistic. No one knows what causes autism, but sometimes children (and adults) who appear bright and normal begin to withdraw and regress until they no longer speak. Perhaps someday sign language-speaking apes can give us some insight into effective ways to communicate with the autistic.

One of the concerns many scientists have had about working with gorillas is that their size and strength make them difficult to handle and potentially dangerous. Patterson knew that Carroll Soo Hoo, who had contributed generously toward purchasing the gorillas for the San Francisco Zoo, often romped with adult gorillas. A trip to the zoo in Basel, Switzerland, where the keeper habitually went into the cage with full-grown females and their little ones, convinced her that the danger was little more than a remnant of the gorilla's inaccurately fearsome reputation, with little basis in reality. There was concern as well that gorillas are difficult, unwilling subjects compared to the extroverted, egotistic, eager-to-show-off chimpanzees. Patterson's experience with Koko has clearly demonstrated that gorillas are willing subjects, to a point. When Koko thinks the lessons have gone on too long or the question she's been asked insults her intelligence, she is perfectly capable of being stubborn, contrary, and bratty. She has enough love of mischief in her that Patterson suspects that some of this occasional refusal to cooperate is for the pure fun of frustrating people.

Once Koko understood what was wanted of her—a phenomenon called "learning to learn"—she made rapid progress. Her periods of greatest learning coincided with those of human children, the period between two-and-a-half and four-and-a-half. During this spurt, the number of signs she knew grew by leaps and bounds and the length of her sentences increased substantially. By the time she was ten years old, Koko used about 500 signs on a regular basis and knew nearly 1,000.

Not everyone at the San Francisco Zoo was as thrilled with Koko's progress as Patterson was. There were serious concerns that Koko was becoming too humanized to be returned to Gorilla World, too used to people to ever live—and most importantly, breed—with other captive gorillas. Many an uncomfortable parley was held to determine Koko's fate. Patterson finally negotiated a compromise: Koko would be removed from the zoo and housed in a trailer on the Stanford University campus and a young male gorilla named Kong, from nearby Marine World, would visit her regularly to keep her enough of a gorilla to breed successfully. The zoo officials agreed, and one foggy morning Koko moved down to Stanford. Unfortunately, Koko and Kong only saw each other about once a month. Kong was getting big and Marine World was finding him hard to handle.

It became clear that Patterson was either going to have to buy Koko or give her back. In fact, buying her was not enough; the San Francisco Zoo wanted her replaced with another young female gorilla. Eventually Patterson purchased a pair of baby gorillas, a male and a female, thinking she could exchange the female for Koko and raise Michael as a companion for Koko. Unfortunately, the little female died shortly after arriving, and the zoo began pressing once again for a replacement. This was the low point of Project Koko, with Patterson ill and heavily in debt, and the zoo intransigent. But by then, Koko had become something of a celebrity and when the local newspapers took up her cause, donations poured in to help purchase her. Public feeling ran so high that the zoo was finally compelled to sell Koko to Patterson. It was 1977, and Koko was safe at last.

Two years later, Patterson was awarded her doctorate in developmental psychology on the basis of her work with Koko, and Koko and Michael were moved to a seven-acre farm in Woodside, California. The move was made on Halloween on the theory that it was the one day of the year when a couple of gorillas in the back seat of the car was least likely to elicit unwelcome attention. The farm was purchased by Patterson and her long-time companion, molecular biologist Ron Cohn, who has the dubious privilege of playing papa to a pair of gorillas. The move coincided with the establishment of The Gorilla Foundation, set up to continue Project Koko and to help improve conditions for gorillas, both in the wild and in captivity. (Memberships may be purchased and donations made to The Gorilla Foundation, Post Office Box 620-530, Woodside, CA 94062.)

In the meantime, the work continues. Now far beyond simple sentences and answering questions, Koko has made it clear that she understands the symbolic nature of language. For one thing, she jokes, and humor requires the ability to understand the norm in order to deviate from it. Koko is not fond of Ron Cohn in his role as the enforcer of good gorilla behavior. She gets back at him by calling him names—*stupid devil* is a favorite—and when she was asked, "What is funny?", Koko replied, "Koko love Ron." The first jokes children make are based on incongruity, on the discrepancy between the listener's expectations and what is said. "The statement was remarkable," Patterson explains, "because at that time Koko almost never used Ron's name sign, preferring to draw on her lexicon of insults when referring to her sometimes stern stepfather." She has on several occasions responded to a question to which the correct answer is *drink* by making the sign on her nose instead of her mouth, which is equivalent to the sign for *rotten*. Koko is also susceptible to a bit of wisely applied reverse psychology. One day when she refused to stop breaking plastic spoons, Ron said, "Good, break the spoons," and Koko stopped at once.

Koko likes practical jokes. She herself is terrified of alligators, although she has never seen one. One of her favorite jokes is to brandish a toy alligator at people. The people are then expected to look terror-stricken and run about screaming. Koko knows people produce the reaction for her benefit, but she still thinks it's wonderfully funny. What is even more wonderful is that a full-grown 250-pound (113 kg) gorilla thinks she needs a toy alligator to scare a human.

Koko has always heard English spoken in conjunction with signing, and she responds to the spoken word as well as to signing. Koko often plays with the sounds of spoken language even though she cannot speak herself, except with sign language. Asked to make a rhyme, she has signed *hair bear* and *all ball* and has rhymed *pink* with *stink*.

Perhaps it is not accurate to say that Koko cannot speak, because she has for a number of years, through a voice synthesizer. When that equipment was left behind in the laboratory at Stanford, Apple Computer company devised a gorilla-proof computer for Koko to use, a modified Macintosh II with a color monitor that gives Koko computer-digitized speech. The words Koko selects from her computer keyboard are spoken in a digitized voice and also appear on the computer screen. Since Koko is now learning to read, this will help reinforce her reading lessons. Koko's reading lessons consist of correctly signing the words on flash cards. Koko reads her name and a number of other words as well.

Michael, too, is proficient in sign language, and the two gorillas sometimes sign to each other, occasionally with signs they have made up between themselves, as children sometimes invent a private language. It is devoutly hoped that Koko and Michael will one day breed. In the meantime, The Gorilla Foundation is looking for land in Hawaii to start a gorilla preserve where the world's first talking gorillas can live and create a family in peace.

It is difficult to measure Koko's IQ (intelligence quotient). Intelligence tests were designed for human children, not gorillas. Answers that would be wrong for a child might be right for a gorilla. For example, asked where she would go in the rain, Koko picked a tree when the correct answer, for a child, was a house. For a gorilla, a tree is a more reasonable response. Testing her is still more difficult because Koko is willing to take tests only so long as they challenge her. Once she has mastered between 70 and 90 percent of the test accurately, she becomes bored and will deliberately give wrong answers until the researchers give up in frustration. She does the same thing with language drills. Once she has mastered the signs to her own satisfaction, she is perfectly capable of giving bratty answers until the grimly determined scientists throw up their hands in despair and let her go do something that's more fun.

Gorillas are intelligent, of that there is no longer any serious question. Koko and Michael have made it undeniably clear that apes can master language, and Koko is beginning to read and master speech with her computer-digitized voice. What then remains to distinguish man from beast, humans from the other apes? It is neither tool-using nor tool-making, intelligence nor symbolic conceptualization, nor the ability to master language.

Science suggests a sad answer. The great apes, our closest relatives, are vegetarian, and while they may bluff and bluster, quarrel and even do battle with each other now and again, their social groups are mainly cooperative. Only man is the killer ape, killing other animals for food and sport, killing each other for territory and personal gain. The world is changing in many ways as people come to a clearer understanding of the interdependence of all living things, plant and animal alike. We can no longer casually render a species like the mountain gorilla extinct simply because the land on which the animals gently roam is wanted to grow pyrethrum. Perhaps the best use of our much-vaunted human intelligence is to find a way to control the human penchant for deadly violence that is the one intractable distinction between us and the great apes.

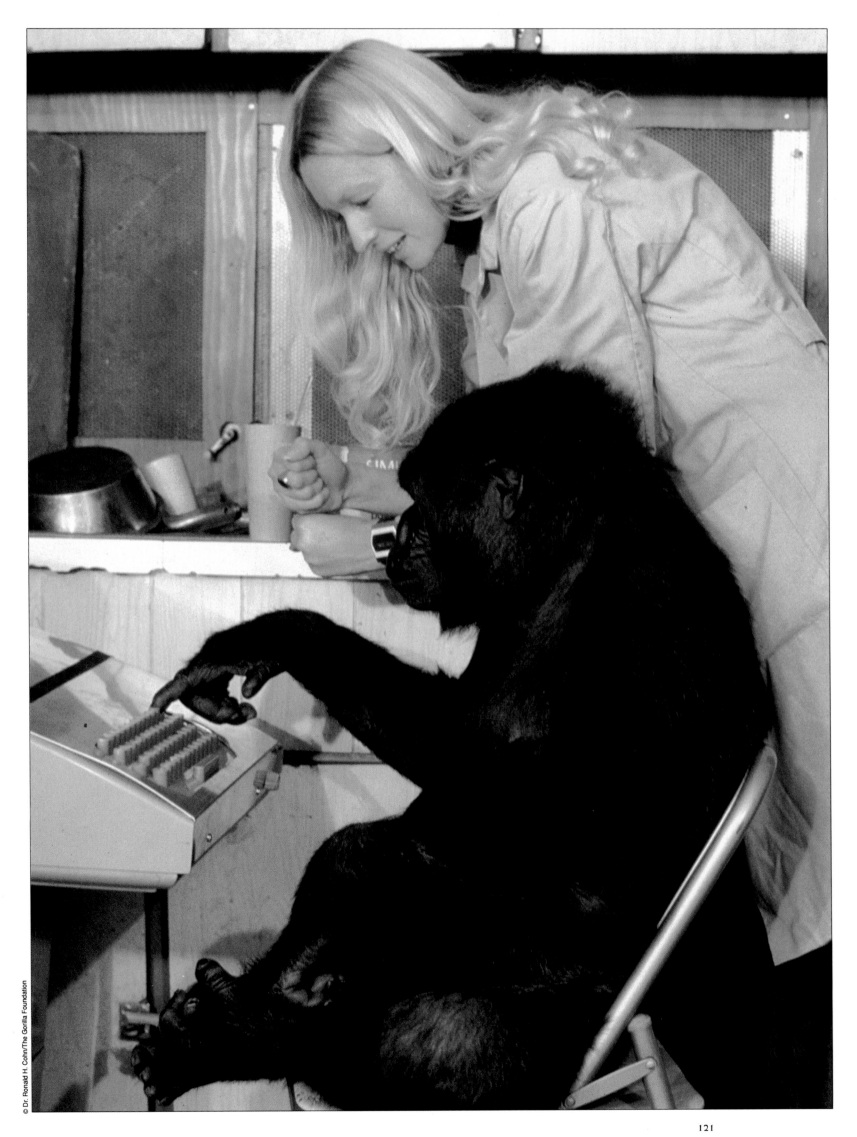

121

Gorilla Foundations

The Digit Fund
c/o The Morris Animal Foundation
45 Inverness Drive East
Englewood, CO 80112

The Gorilla Foundation
P.O. Box 620-530
Woodside, CA 94062

The Wildlife Preservation Trust International
34th Street and Girard Avenue
Philadelphia, PA 19104

Mountain Gorilla Project
c/o African Wildlife Foundation
1717 Massachusetts Avenue, Suite 602
Washington, DC 20036
or
Craig P. Sholley, Director
Mountain Gorilla Project
B. P. 36
Ruhengeri, Rwanda

World Wildlife Fund
1250 Twenty-Fourth St. NW
Washington, DC 20037

World Wildlife Fund UK
Panda House
Weyside Park
Godalming
Surrey
England GU7 1XR

Bibliography

Akeley, Carl. *In Brightest Africa.* New York: Doubleday, Page, & Co., 1920.

Barnum, P. T. *The Wild Beasts, Birds, and Reptiles of the World: The Story of their Capture.* 1888.

DeVore, Irven (ed.). *Primate Behavior.* New York: Holt, Rinehart, & Winston, 1965.

Durrell, Gerald. *The Stationary Ark.* New York: Simon & Schuster, 1976.

Fossey, Dian. *Gorillas in the Mist.* Boston: Houghton Mifflin Company, 1983.

Green, Susan. *Gentle Gorilla: The Story of Patty Cake.* New York: Richard Marek Publishers, 1978.

Hahn, Emily. *Eve & the Apes.* New York: Weidenfeld & Nicolson, 1988.

Hallet, Jean-Pierre with Alex Pelle. *Animal Kitabu.* New York: Random House, 1967.

Hamburg, David A. and Elizabeth R. McCown (eds.). *The Great Apes.* Menlo Park, California: Benjamin/Cummings Publishing Co., 1979.

Johnson, Martin. *Camera Trails in Africa.* New York: Grosset & Dunlap, 1924.

Johnson, Osa. *I Married Adventure.* New York: J. B. Lippincott Company, 1940.

Morris, Desmond. *Animal Days.* New York: William Morrow and Company, 1979.

Morris, Desmond. *The Naked Ape.* New York: McGraw-Hill Book Co., 1967.

Mowat, Farley. *Woman in the Mists.* New York: Warner Books, 1987.

Nichols, Michael. *Gorilla: The Struggle for Survival in the Virungas.* New York: Aperture, 1989.

Patterson, Francine and Eugene Linden. *The Education of Koko.* New York: Holt, Rinehart, & Winston, 1981.

Sanderson, Ivan T. *Animal Treasure.* New York: The Viking Press, 1937.

Schaller, George. *The Year of the Gorilla.* New York: Ballantine Books, 1964.

Simonds, Paul E. *The Social Primates.* New York: Harper & Row, 1974.

Wolpoff, Milford H. *Paleo-Anthropology.* New York: Alfred A. Knopf, 1980.

Index

Page numbers in italics refer to captions and illustrations.